MICROPHONE IN HAND

Interviews, Forewords, Reviews:
Revelations from a Scientist and about a Scientist

MICROPHONE IN HAND

Interviews, Forewords, Reviews:
Revelations from a Scientist and about a Scientist

Istvan Hargittai

World Scientific

NEW JERSEY · LONDON · SINGAPORE · BEIJING · SHANGHAI · TAIPEI · CHENNAI

Published by

World Scientific Publishing Co. Pte. Ltd.

5 Toh Tuck Link, Singapore 596224

USA office: 27 Warren Street, Suite 401-402, Hackensack, NJ 07601

UK office: 57 Shelton Street, Covent Garden, London WC2H 9HE

British Library Cataloguing-in-Publication Data
A catalogue record for this book is available from the British Library.

MICROPHONE IN HAND
Interviews, Forewords, Reviews: Revelations from a Scientist and about a Scientist

ISBN 978-981-98-0921-9 (hardcover)
ISBN 978-981-98-0922-6 (ebook for institutions)
ISBN 978-981-98-0923-3 (ebook for individuals)

For any available supplementary material, please visit
https://www.worldscientific.com/worldscibooks/10.1142/14212#t=suppl

Desk Editor: Rhaimie Wahap

Typeset by Stallion Press
Email: enquiries@stallionpress.com

For Magdi

ABOUT *MICROPHONE IN HAND*

Istvan Hargittai has taught us how symmetry is all around us, led us on walking tours of the great scientific cities of the world, interviewed many Nobel Laureates (to let us see just how normal human beings they are), and taught us of good Russian science and what was special about the Martians who were not from Mars. In this readable book, he turns his microphone around, so to speak, and allows us to see and understand the man. Through personal interviews, prompted by others to tell his story. It is in fact many stories, of an engaged life of a 20th-century scientist and writer, moving through political curtains. Aware of the details of life, people, sensitive to the good in them, cognizant of the politics swirling around them. A life worth recounting, a friend worth having — you will like Istvan Hargittai's book.

— Roald Hoffmann, Chemist, Nobel laureate, and writer

I admire Istvan Hargittai's books, often with his wife Magdolna, about Nobel laureates and other accomplished scientists and the conditions under which they worked, some comfortable, others quite the opposite. Here are fascinating thoughts about science and society from a true intellectual.

— P. James E. Peebles, Physicist, Nobel laureate,
Princeton University

Istvan Hargittai offers a piece of history and history of science in this volume. Exciting reading what I recommend to everybody who is interested in history and science and wants to know about and interesting personality.

— Ivan T. Berend, Historian,
Distinguished Research Professor, UCLA,
Former President of the Hungarian Academy of Sciences

This erudite and charming collection of interviews and articles reflect the writing of Istvan Hargittai over 40 years, commenting on and illuminating science and how science interacts with society. Read this and be informed and entertained about the international activities of science and scientists and their importance for the future.

— Paul Nurse, OM CH FRS,
Nobel laureate in Physiology or Medicine,
Director of the Francis Crick Institute (London),
former President of the Royal Society

To some, the pursuit of scientific discovery may appear to be a detached, solitary endeavor, devoid of societal influences, cultural biases, and meaningful human interaction. However, in a compelling series of interviews with leading scientists from around the globe, Istvan Hargittai has revealed in a series of past publications the rich human drama behind the quest to unlock nature's secrets, effectively challenging this misconception. The exhilaration and anguish that accompany the birth of new knowledge can be a lonely journey, yet it unfolds against the backdrop of societal pressures. In this work, the focus shifts to Hargittai himself, allowing readers to discover his remarkable life and career. This book is a compelling read for anyone seeking to gain a deeper understanding of the scientific enterprise.

— Richard N. Zare, Marguerite Blake Wilbur
Professor of Natural Science,
Stanford University, Wolf Prize, King Faisal Prize

Istvan Hargittai is one of those rare, curious individuals who knows that science is made by scientists, and their life stories often explain the unpredictable roads to discoveries. Hargittai is a good listener and an extraordinary storyteller who interviewed numerous world-class scientists in various disciplines, from Edward Teller to Francis Crick. Many exciting details often did not fit into the interviews that served particular goals and were restricted by word limits. This time, he is the "victim" of the microphone, and we learn about the fascinating events left out of those interviews and the circumstances of how he was able to "hunt down" his heroes. But perhaps more importantly, here the scientist, writer, and family man are exposed, reflecting the lives of many like-minded individuals who struggled through the tumultuous times of the communist era in Hungary. If you want to learn about the life and science of the historian and structural chemist Istvan Hargittai, this volume is your chance.

— Gyorgy Buzsaki, Biggs Professor of Neuroscience,
New York University
Grossman School of Medicine, Brain Prize,
author of *The Brain from Inside Out*

ABOUT THE AUTHOR

Istvan Hargittai is a physical chemist and professor emeritus (active) at the Budapest University of Technology and Economics. He is a member of the Hungarian Academy of Sciences and the Academia Europaea (London), and a foreign member of the Norwegian Academy of Science and Letters (Oslo). He is a PhD and DSc, and has honorary doctorates from Lomonosov Moscow State University, the University of North Carolina, and the Russian Academy of Sciences. He has been Founding Editor-in-Chief of the international journal *Structural Chemistry* (Springer Nature) since 1989. He has published over three hundred research papers and reviews and has published hundreds of other papers, in particular, in science dissemination and popularization. He has authored and edited over 50 books about structural chemistry, history of science, the nature of scientific discovery, memorials of scientists, conversations with famous scientists, and other topics. His books have appeared in English, Hungarian, Russian, German, Swedish, Italian, Japanese, Chinese, and in the Farsi language.

ALSO BY THE AUTHOR

With Balazs Hargittai, *Brilliance in Exile: Diaspora of Hungarian Scientists from John von Neumann to Katalin Karikó* (Central European University Press 2023).

With Balazs Hargittai, *Quotable John von Neumann: Thoughts of a Great Scientist* (Neumann Society 2023).

Open Government, Open Diplomacy: Conversations with a Former American Diplomat in Budapest, André Goodfriend (Central European University Press 2023).

With Magdolna Hargittai, *Science in London: A Guide to Memorials* (Springer Nature 2021).

Mosaic of a Scientific Life (Springer Nature, 2020).

With Magdolna Hargittai, *Moscow Scientific: Memorials of a Research Empire* (World Scientific, 2019).

Edited by Balazs Hargittai, *Culture and Art of Scientific Discoveries: A Selection of István Hargittai's Writings* (Springer Nature, 2019).

With Magdolna Hargittai, *New York Scientific: A Culture of Inquiry, Knowledge, and Learning* (Oxford University Press, 2017).

With Balazs Hargittai, *Wisdom of the Martians of Science: In Their Own Words with Commentaries* (World Scientific, 2016).

With Magdolna Hargittai, *Budapest Scientific: A Guidebook* (Oxford University Press, 2015).

With Balazs Hargittai and Magdolna Hargittai, *Great Minds: Reflections of 111 Top Scientists* (Oxford University Press, 2014).

Buried Glory: Portraits of Soviet Scientists (Oxford University Press, 2013).

Drive and Curiosity: What Fuels the Passion for Science (Prometheus, 2011).

Judging Edward Teller: A Closer Look at One of the Most Influential Scientists of the Twentieth Century (Prometheus, 2010).

With Magdolna Hargittai, *Symmetry through the Eyes of a Chemist* (3rd Edition, Springer, 2009, 2010).

With Magdolna Hargittai, *Visual Symmetry* (World Scientific, 2009).

The DNA Doctor: Candid Conversations with James D. Watson (World Scientific, 2007).

The Martians of Science: Five Physicists Who Changed the Twentieth Century (Oxford University Press, 2006, 2008).

The Tragedy of Edward Teller (Hungarian Academy of Sciences, 2005).

Our Lives: Encounters of a Scientist (Akadémiai Kiadó, 2004).

The Road to Stockholm: Nobel Prizes, Science, and Scientists (Oxford University Press, 2002, 2003).

With Balazs Hargittai and Magdolna Hargittai, *Candid Science I–VI: Conversations with Famous Scientists* (Imperial College Press, 2000–2006).

With Magdolna Hargittai, *In Our Own Image: Personal Symmetry in Discovery* (Kluwer/Plenum, 2000; Springer, 2012).

With Magdolna Hargittai, *Symmetry: A Unifying Concept* (Shelter, 1994; Random House, 1996).

FOREWORD

Microphone in Hand by Istvan Hargittai provides a window on the inner workings of science and scientists. It presents signposts and pointers to his earlier books, which include collections of interviews and conversations with many prominent scientists plus books with scientific themes such as Symmetry, DNA, or Nobel Prizes. There are also geographically focused collections centered on New York, London, and Moscow, places where Istvan and Magdolna have worked.

I first encountered Istvan and his scientific partner Magdolna through their work on electron diffraction of gases, driven by my own interests in X-ray and electron diffraction. Their 1988 book provided a solid foundation for the "Stereochemical Applications of Gas Phase Electron Diffraction". Their interests expanded over many years to cover a broad portfolio of research areas: a consuming curiosity about science, scientists, scientific cultures, and what factors catalyze successful discoveries and inventions. They are very interested in the way scientists work, and the importance of scientists in the wider world even though they make up only 0.2% of society.

The book is in two parts. The first part contains interviews with Istvan that provide insights into the climate of research, education, and society during his upbringing and career in Hungary and the wider world, including his experiences as a child during the wartime persecution of Jewish people and Jewish scientists. The second part reproduces excerpts from Forewords and Book Reviews of Istvan's published books often by scientists with great insight. "Microphone in Hand" has turned out to be a novel way of creating a biographical/autobiographical fusion of Istvan's life and philosophy.

It includes many interesting vignettes, including a superb educational interview with Lou Massa on Symmetry, observations about the rate of chiral center inversion in the drug thalidomide that underpins its therapeutic and teratogenic effects, as well as insights into quasicrystals and noble gases. There is strong advocacy of the talent and achievements of Hungarian scientists like Wigner, Szilard, and Teller, plus Istvan's own advice for a young scientist to "find an activity which they are good at and can excel". The book is a treasure of interesting anecdotes, reported in an honest and perceptive way.

Richard Henderson
CH, FRS, Biophysicist,
former director of the MRC Laboratory of
Molecular Biology, Cambridge,
Copley Medal (Royal Society), Nobel Prize

CONTENTS

INTRODUCTION

"The best one can do is doing what one is best at doing."
(*Drive and Curiosity*, 2011)

One of my favorite books is a small collection of interviews in which Oliver Sacks (1933–2015), a British-American neurologist and author of popular science books, was the interviewee.[1] Years ago, our common interest in interviewing brought Sacks and me together. Between 1990 and the early 2000s, I recorded and published conversations with many famous scientists; most of them were published in the six-volume *Candid Science* book series.[2] Sacks was not among my interviewees. We met in 2003 and stayed in close contact for a few years. He was well acquainted with the *Candid Science* volumes. We corresponded intensively by old-fashioned hard copy letters as Sacks did not use e-mail. We both felt that any interview characterized both the interviewee and the interviewer. A good interview, even if it is short and focuses only on a few questions, can give a complex picture, however impressionistic, of both personalities.

Sacks' example inspired me to compile a selection of interviews that had been recorded with me over the years. I communicate 22 conversations or excerpts from them, with minimal editorial

---

[1] Oliver Sacks, *The Last Interview and Other Conversations*. Brooklyn and London: Melville House, 2016.

[2] B. Hargittai, M. Hargittai, I. Hargittai, *Candid Science I–VI*. London: Imperial College Press, 2000–2006.

Istvan Hargittai and Oliver Sacks (right), 2003, at a New York reception, by Magdolna Hargittai.

changes, supplemented by photographs and footnotes where appropriate.

While looking for these interviews in the literature, I came across reviews of my books and felt that they complement the interviews. From there, it was just a step to include a selection of forewords that others had written for my books. I thought it more interesting to present several reviews of the same book than to use the available space to reproduce reviews of as many books as possible. Among the books figuring in this compendium, several were produced jointly with my wife, Magdolna, or Magdi for short, and with our son, Balazs.

To me, the main strength of this book is that it covers four decades, from the early 1980s to the early 2020s. In addition to the personal perspective, this period spans several political systems in Hungary, from the hesitantly relaxing socialist dictatorship via short-lived democratic governance in the 1990s to yet another autocracy that has gradually dismantled democratic institutions since 2010. During these four decades, my wife and I did considerable traveling and were often visiting professors, so several of the interviews with me had an international flavor.

My life started in 1941 in Hungary under the pro-Nazi and increasingly harsh anti-Semitic Horthy regime, which in 1944 put me on a train of cattle carriages destined for Auschwitz, to certain death. My father had already been murdered in 1942 in the so-called forced labor service. Due to a series of coincidences, we — my mother, 10-year-old brother, and I — ended up in a ruthless labor camp in Vienna rather than Auschwitz. We lost everything but survived.

I was driven by the joy of learning from an early age. The so-called socialist regime in Hungary tried to prevent me from studying when, in 1955, they refused to let me go to high school in the same town from where we had been deported in 1944. In 1959 they tried to prevent me from going to university. I was finally allowed to study, first at Eötvös Loránd University and then at Lomonosov Moscow State University. On my return home in 1965, I started building my research group and a new field of research as an associate of the Hungarian Academy of Sciences. Then, the change of the political regime in 1989–1990 opened my way to a university professorship. The interviews with me reflect this life journey.

This book is my testimony to the unity of the two cultures and to the internationalism of science. It presents my personal experiences as a scientist, and I hope it resonates with the personal experiences of many readers.

Istvan Hargittai
Budapest, Fall 2024

Note on permissions to reproduce published material. I express special thanks to publishers and individuals for permission to reproduce interviews and reviews. Alas, in some cases the fee for permission to reproduce was unaffordably high for me. In a few other cases, the permission was impossible to obtain for other reasons. I still listed the bibliographic information of the reviews with a brief quote from them. I have made every effort to locate the copyright owners and ask for permission. I apologize if any permission to reproduce I may have improperly stated or overlooked, and if notified will make amends in later printings.

Part One

Selection of Interviews

I

ZSUZSA REGŐS, "RESEARCHING MOLECULES," CONVERSATION WITH PROFESSOR ISTVAN HARGITTAI

Vengerskie Novosti 1981, 3, 16–17 (former Russian-language *Hungarian* magazine)

What do you do?

My seven-year-old daughter thinks I'm a *molecule doctor.* I have a doctorate in chemistry, I head the structural research department of the natural science laboratories of the Hungarian Academy of Sciences, and I have a title of Professor at Eötvös University. But the titles are not important. The work is interesting. We use electron beams to investigate the structure of molecules.

So, you're a chemist. How did you choose this as a profession?

Math was my favorite subject as a child. In the fifth grade of primary school, I won a math competition, and my prize was a chemistry book. It fascinated me so much that I decided to become a chemist. I still have the book and looked at it the other day. I was surprised by how much of it was about, economics, foreign trade, etc., and most

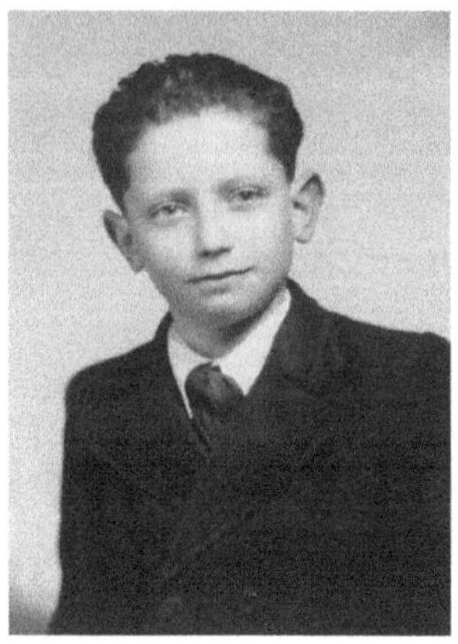

Eleven-year-old Istvan Hargittai, by unknown photographer, and his first chemistry book.

of it was communist propaganda, but for me, it was just about chemistry.

You completed your studies in Moscow. How did it happen?

I started at Eötvös University and already as a freshman I started doing some rudimentary research which was not very exciting. When I was a sophomore, I heard about study abroad opportunities and applied, without knowing exactly what I was doing. I was accepted. I knew the Russian grammar well enough, but I was nowhere in everyday language. In the summer of 1961, I attended a three-week language course. I continued on in Moscow as a second-year student.

Was it hard?

The first few months. But I tried to practice my language skills as much as I could, made friends with Russian students, and immersed myself in the life there. Some of my fellow Hungarian students complained that I was neglecting the Hungarian community. I didn't use a dictionary, I tried to understand everything by listening to Russian speech. What I regret is that I did not have time to get to know the country better. At the end of each academic year, I always rushed home.

And following graduation?

In Moscow, I was introduced to an area of research that was new in Hungary, the investigation of molecular structure by electron diffraction. I found a place where I could initiate such a research direction, and I am still there. At that time, it was a small independent research laboratory of the Hungarian Academy of Sciences whose director was looking for new research directions. Our ambitions were in sync.

With József Hernádi (right), who oversaw the machine shop, 1966, by unknown photographer.

I owe a lot to József Hernádi, who oversaw the machine shop. Sadly, he is no longer with us. He gave me invaluable help. The workshop played a crucial role in enabling me to start my new research direction because I needed to build new equipment with the right tools and imaginative expertise. It may sound too romantic, but this is what happened as Hernádi told me years later. He had read a novel by Lajos Szilvási about a young engineer who had graduated in the Soviet Union and started work in a plant where he had to overcome many difficulties, and then an old master came to his aid. This is how he viewed our situation. He was an excellent man and had special talent. Together we published articles in international journals.

How did you progress in your work?

With Lev V. Vilkov (left) in the Moscow Kreml, 1970s, by unknown photographer.

I stayed in contact with my professor in Moscow, Lev V. Vilkov. We corresponded in detail, I asked him questions, which he answered, and later he visited our laboratory. In time, his students came to study with us in Budapest.

In Budapest, my international outlook broadened, and we invited the famous Norwegian professor, Otto Bastiansen. He was an energetic personality and gave an excellent lecture at the Academy of Sciences, which was a great boost for our field of science. He invited

With Otto Basiansen (left) in Oslo, 1969, by unknown photographer.

me to Norway for three months. During my stay in Oslo, Bastiansen was invited to be a visiting professor in America. The invitation included the possibility for him to take two young colleagues with him and he chose me as one of them. I spent a year at the University of Texas at Austin and made very good contacts there. Many Americans then came to us for extended visits, and I went to the United States to give lectures.

At my latest lecture in Moscow, in the autumn of 1980, I did something innovative. As a hobby, I had collected images of the occurrence of symmetry in nature and in our surroundings, which I could relate to our science. At the end of my presentation, I showed a selection of my symmetry images. This was very well received. I have always been attracted by the challenge of promoting science, and symmetry has proved to be a useful tool for doing so. You obviously know what symmetry is.

Of course.

Then you are lucky because I have never been able to explain it unambiguously. It happens everywhere, in life, in science. We use it all the time in scientific research. Although it is the most general concept, it is difficult to define it precisely. Its most common kind is what even children notice, it's mirror symmetry. Our own children have also made some drawings for my book on symmetry.

How involved is your wife in your work?

She is a chemist too, we met at university. I infected her with my interests, but she started with her own research direction. We have written a book together, which was translated into English and Russian.

What have you been doing lately?

There is a new aspect to our work, in which our son, Balazs, played an indirect role when he was 5 years old. In 1976, Balazs and I spent some time in Mátrafüred, a resort of the Academy of Sciences. It was the last year before Balazs started school. There, he played together with a little girl of the same age, Aliz, who was there with her father. We, the two fathers, also got to know each other. Aliz's father was a staff member at the Debrecen Institute for Nuclear Research and our conversations developed into a collaboration. We combined our two experimental techniques, electron diffraction and quadrupole mass spectrometry. The new, combined experiment allowed us to investigate the structure of molecules that were unstable. It would have been impossible for us to investigate without monitoring the composition of the sample during the experiment (this is what the mass spectrometer did). This new direction of research has attracted international attention.

2

B.T., "PRIZE FROM AMERICA" (EXCERPTS) *ÖTLET* (FORMER HUNGARIAN MAGAZINE OF INNOVATIONS) 1987, SEPTEMBER, P. 30

The occasion for the interview is that Istvan Hargittai edited an interdisciplinary magazine issue on symmetry. It was about a thousand pages in the journal *Computers and Mathematics with Applications*. The Association of American Publishers awarded this contribution in 1987 with the recognition of the best journal issue in the preceding year.

How did you come to edit the award-winning issue?

Back in 1983, I published a small Hungarian book whose title in English is *Symmetry through the eyes of a chemist*. Ervin Y. Rodin, professor of mathematics at Washington University in Saint Louis, Missouri, happened to come across this book.[1] His mother tongue is

---

[1] Ervin Y. Rodin immigrated in the United States from Hungary after the 1956 revolution. He studied at the University of Texas at Austin, majored and acquired his PhD in mathematics. His research was in large-scale optimization and artificial intelligence. At the time we were in contact, during the second half of the 1980s, he was Professor of Electrical & Systems Engineering at the McKelvey School of Engineering, Washington University, St. Louis, Missouri. We never met in person.

Hungarian, so he had no language difficulties. In 1983–1985, I was a visiting professor at the University of Connecticut. He called me on the phone and invited me to compile a special collection of articles on symmetry in one of his journals.

I pretty much had a list of authors, as I had met a lot of interesting people at various conferences and universities. Then the list of authors grew, and many people recommended someone or contacted me to contribute an article to the special issue. I invited potential authors, most were enthusiastic and, of course, I also met with rejections.

With Kurt Mislow (left) of Princeton University at a symmetry meeting in Darmstadt, Germany, 1986. Mislow holds the first large, edited volume of *Symmetry*, by unknown photographer.

I approached one of the famous scientists at Princeton University, but he evaded me, saying that although he found my idea very interesting, he would not publish it in such a relatively unknown journal. Then we met at a symmetry meeting in Darmstadt, Germany in 1986, right after the first big *Symmetry* volume had appeared to which he declined to contribute. It was a great satisfaction for me when Mislow congratulated me on the publication of this volume. By then, I was already working on its sequel.[2]

You mentioned that Professor Rodin first contacted you when you were at the University of Connecticut. How long have you been lecturing in the U.S. and how do you find the education system there?

I first spent a year at the University of Texas at Austin in 1969. After that, I went to several places and have been teaching in American universities for about half of the last five years. It's difficult to give a brief description of the U.S. education system because it is very diverse and flexible. It is often criticized here in Hungary for being of low quality. This is partly true. It is true that you can graduate from a university without ever studying physics or chemistry if you are not interested in them. But if a young person is motivated, he or she can enroll in a high school [i.e., secondary school] where the subjects of interest are taught at the university level. In America, they do not try to adjust the level of education to the average. Many people can get into universities, which can be of very different levels. In Hungary, the freshmen are generally better prepared than those from most of the American high schools, but the student population at the Hungarian university is highly selective, a relatively narrow group. To have a system of higher education that offers a choice of a broad variety of levels as it is in the United States, a great deal of resources is needed.

---

[2]The two volumes were published after their materials had first appeared as journal issues: I. Hargittai, Ed., *Symmetry: Unifying Human Understanding* (Pergamon Press, 1986) and I. Hargittai, Ed., *Symmetry 2: Unifying Human Understanding* (Pergamon Press, 1989).

3

INTERVIEW BY LOU MASSA, CITY UNIVERSITY NEW YORK (CUNY) TELEVISION, MANHATTAN, 1997

The interview was recorded and broadcast in the framework of Professor Massa's "Science and the Written Word" Program. The topic was the book of Istvan Hargittai and Magdolna Hargittai, *Symmetry: A Unifying Concept* (Bolinas, California: Shelter, 1994). The transcripts are reproduced with permission from Lou Massa.

Let's begin with the title of your book. Could you give me a rough idea of its meaning?

Symmetry is one of those terms, which may be looked at as a technical term, but it is also part of our everyday language. This gives us the benefit of not having to explain it every time we talk about it. On the other hand, it is very difficult to give an exact definition for it. Symmetry is characteristic of an object or even a phenomenon if we can say that if we do something to this object or phenomenon and after we have done this, it reappears the same way as it was before. An example is a mirror line in which we reflect one half of our face in it and our face reappears as if nothing had happened to it.

So, it is an invariance associated with an operation on an object.

What you say is a more mathematical approach. The way I said was a more everyday-like approach. One of the greatest mathematicians of

Istvan Hargittai with Lou Massa (right) in the studio of CUNY TV, 1997.

the twentieth century, Hermann Weyl, also described symmetry like the way I just did. Of course, the rigorous way is how you did it.

Can you break down symmetry into various classes of its appearance?

Before that, I return to your initial question: Why is it a unifying concept? It is a very general and fundamental concept, both in our life, in science, in nature, and it is present everywhere, it can be used for linking and, thus, for unifying. In school, we learn everything by subjects, we learn separately about biology, chemistry, physics, history, human behavior, and so on.

Symmetry, as a general concept, helps bringing back the unity of everything because it is so omnipresent. Once you consider symmetry, you will start relating everything that you see to it, while walking in the woods and looking at plants and while walking in a city and looking at buildings. It is a great bridging concept. It can make connections.

So, virtually, anywhere you have structure, these ideas of symmetry will arise in a natural way. Descartes was fond of saying that a circle, which is microscopic, has exactly the same properties as a circle, which is macroscopic. That's another sort of unification. That is to say that symmetry applies at every scale of distance.

I can make two comments in this connection. The symmetry of the very small and the very large circle is the same, they differ only in their size; everything else is the same. We call this similarity symmetry. This makes it possible for us to discuss a series of related things, which are not identical but similar. Another side of the consideration of scale is when you look at something from a distance, you may see that it is symmetrical and then you get closer and you notice irregularities, the symmetry is no longer perfect. That is another consideration of scale. It is up to us, it is our decision, how rigorous we want to be considering something symmetrical or not. If we are magnanimous rather than rigorous, we will find the symmetry concept to be applicable to a much broader range of objects and phenomena than a more restricted approach would allow. Geometrical symmetry is rigorous and confining. In real life and in most of science, the more magnanimous approach is most fruitful, and this is what we represent in this book.

What about the classes of symmetry?

There are two major kinds. The technical terms are point groups and space groups. In everyday language, in the case of a point group during the symmetry operation, during the change we are performing, at least one point remains unchanged.

Examples of a point group (left) and a space group (right) in typical Italian pavement patterns (photographs by Istvan Hargittai).

Imagine a circular pattern, for example. You can rotate it, at least in your imagination and the center point does not move, it remains where it was. The whole pattern keeps reappearing during rotation while the pattern is moving around, except for the center point which stays in place. The same happens when you reflect the pattern by placing a mirror through the center point and the left turns into the right or the right turns into the left, thereby recreating the whole pattern in the process of reflection. In this case, during this operation, not only the center point but all points along the mirror, the reflection plane, stay in place and remain unchanged. In point group symmetry, at least one point (during rotation) or more points (during reflection) stay in place.

At least one point is invariant to all operations.

Correct. In the case of space group symmetry, there is not even a single point that would be left alone during the symmetry operation. Such a pattern extends, at least in our imagination, to infinity, it is being repeated and repeated endlessly. To remain with the pavement example, in this case, the pattern is being repeated endlessly. In physical reality, there is no such infinite pavement, but there easily is in our imagination. In this case, a mere repetition is the simplest symmetry operation, which creates such a pattern. This is also a very common kind of symmetry. Usually, when people think about symmetry, this is not what comes to mind. Wallpaper patterns have this kind of symmetry.

For point group symmetries, the operations are reflection, rotation, inversion. What are the elements of space group operations?

The simplest is translation. Shift the motif by a certain distance and never stop doing this.

For a wallpaper…

You choose a motif, and you translate by a fixed distance, and you keep doing this and create the pattern. There are then other ways, other operations, to create a pattern.

Footprints.

In addition to simple translation, let's mention one more operation for space group symmetries. You leave your footprints while walking on wet sand. While walking, there is a left footprint, then a right footprint, and so on. The footprints are not being repeated by simple translation because the left footprint and the right footprint are different. Imagine a line along the footprints, you start with a left footprint and shift it to a certain distance and reflect it, using the imaginary line for reflection. This way the right footprint is created. Then you shift this right footprint to the same distance as before, reflect it and there is another left footprint, and so on. Translation and reflection are being applied consecutively. This operation is called glide reflection, and it is another frequent pattern on wallpapers and elsewhere.

Let's talk about specific examples. Bilateral symmetry. Is it everywhere?

This is the most common example of symmetry. Our face has bilateral symmetry and so has our body with some simplification.

Which is also called mirror symmetry.

Of course. We must be magnanimous because nobody has a perfect bilateral symmetrical face. Just as well because it might be dull. We should not be too perfect but should not be too imperfect either. Some imperfections of the face are sometimes called "sex appeal." Speaking about the widespread occurrence of bilateral symmetry, we like to speak about functionality. Why do we have bilateral symmetry? We walk along the surface of the earth. We don't move vertically, and our top and our bottom are different. We move forward and not backward, and our front and our behind are different. But turning left and turning right is about the same and our left and right are equivalent. Not quite but close. This is the functionality of bilateral symmetry. At least, this is how we explain it. This is the symmetry insects and animals have and this is why we build our cars and airplanes to have bilateral symmetry. The Lunar Module does not have bilateral symmetry because its main mode of motion is vertical rather than horizontal.

The butterflies have elaborate patterns with perfect symmetry. As you indicated, the human form can be closely symmetrical along the middle line of our body. Animals and anything that moves on the surface of the earth have an advantage associated with bilateral symmetry. This is ubiquitous. It seems that mankind has also incorporated such symmetry in all cultural, scientific, and architectural endeavors. It seems to have great aesthetic appeal as well.

There is also a degree of symmetry that we seem to feel comfortable with. Too much symmetry might be too much. If you have a highly symmetrical building, it may be confusing to find your way in it.

What about rotational symmetry? What comes to mind about its appearance in nature?

Many flowers have mirror symmetries, just bilateral, as orchids, or multilateral, as many others. Multilateral symmetries occur when there is a combination of reflection and rotation. More seldom, though not infrequent in flowers, is rotational symmetry only. Rotational symmetry means that during rotation for a complete circle, the original object reappears twice (twofold rotation), three times (threefold), and so on. There is no limit in this. In human creations, like windmills and pinwheels, there is fourfold rotational symmetry. Hubcaps on cars may have reflection and rotation, and may have rotation only. To my sense, rotational symmetry only is more consistent with motion than when reflection, that is, mirror symmetry is involved. Rotation only induces the feeling of motion. Reflection induces the feeling of stopping motion. Propellers have only rotational symmetry. Considering company logos, symmetries may express functionality. There are many railway companies in Europe; every country has its railway company, and there may be more. Most of their logos have twofold rotational symmetry. Such a logo may indicate that you go somewhere and come back. A symmetry plane would induce the feeling of stopping. The logos of banks may also relate to functionality even if it is more intricate. Some banks have fourfold rotational symmetry with no reflection. Does this indicate that they move around, they rotate money? Maybe.

So, symmetry has deep psychological aspects.

It may.

Let me bring your attention back to a symmetry associated with the relation between the left and right hands.

A very important aspect. None of our hands is symmetrical. But the pair of our hands together makes a symmetrical system. There is a mirror plane between our left and right hands, but the two hands are impossible to superimpose. This kind of peculiar symmetry is called handedness — chirality — and the name comes from the Greek word for hand.

You have two objects, which are images of one another.

Mirror images.

And they are not possible to superimpose them. Is it important?

Very important. It has been known for a long time, but in science, the headway was made by Louis Pasteur in 1848 when he observed crystals of the same substance that came in two versions, left-handed and right-handed. The assignment of the names was arbitrary, which was the left and which was the right. The two versions were each other's mirror images but not superimposable. This observation initiated a whole new direction in chemistry. Chirality is vitally important.

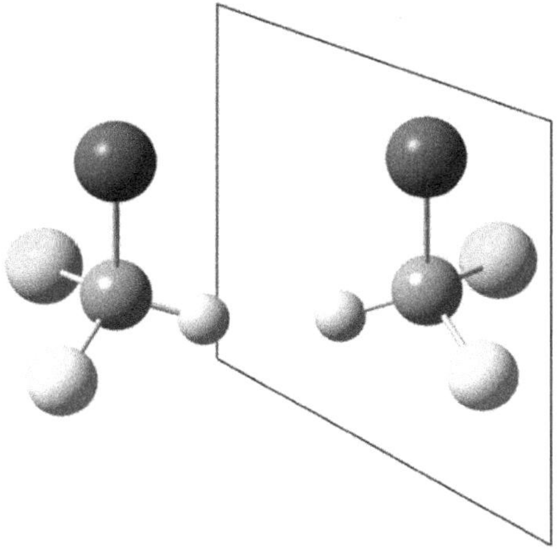

Model of a molecule and its nonsuperimposable mirror image.

Molecules of many substances come in handed versions. A simple example is having a carbon atom with four different atoms linked to it. If there is a substance whose molecules include such segments, they will exist in two versions. The two can be difficult to distinguish, may have many identical properties, yet may have drastically different physiological effects. One of the two chiral versions of Asparagine tastes bitter and the other version tastes sweet.

And they are related as our left and right hands.

I dream about a sugar that would taste sweet but would have the wrong-handedness to be built into the human organism and cause obesity.

It would just pass through.

There are some tragic examples. Thalidomide was a drug in Europe under the name of Contergan in the 1950s. Pregnant women took it against morning sickness. Eventually, it turned out that it caused severe birth defects. The substance had left-handed and right-handed versions; one mitigated morning sickness, but the other caused birth defects.

We were very lucky in the U.S. because we banned its use in our country.

You did not ban it. A conscious associate of the Food and Drug Administration (FDA) kept initiating additional examinations of the validity of the claims about it, so it did not get the necessary green light for marketing in the United States. The FDA was heavily criticized for being slow and holding up the process, but this slowness proved beneficial. The tragic case of Contergan accelerated new legislation of introducing new rules for chiral substances in medicine. The chiral variation must be resolved and only the right chiral version can be marketed for treatment. The chiral separation has become a huge industry in pharmaceutical production.

Let's go back to rotational symmetries. There are an infinite number of polygons and there is an infinite number of rotational symmetries. For regular polygons, there are the triangle, square, pentagon, up to infinity.

At infinity, there is the circle. That is the limiting case. There we have an infinite-fold rotational symmetry. No matter how little you turn around the circle, you still have the same circle again.

However, it is fascinating that when you come to three dimensions, there are no infinite number of regular polyhedra.

Far from it. Let's start with definitions. A regular polygon has equal sides and equal interior angles. Regular hexagon characterizes the snowflake. It is also a puzzle when we try to understand how these regular, symmetrical shapes are created. Johannes Kepler, as early as the beginning of the seventeenth century, was looking at the snowflakes, and he was the first ever who came to the notion that it must be the internal structure — today we would say the network of water molecules defined by relatively weak hydrogen bonds — that determines their external shapes. Kepler came to this notion without knowing about molecules and hydrogen bonds, and intermolecular forces, but he had tremendous intuition. He sensed the importance of packing in creating shapes. Over the centuries, some of the best minds were trying to understand snowflake formation and we still don't fully understand it but have learned a great deal about it. An American photographer Wilson A. Bentley in Vermont collected thousands and thousands of photographs of snowflakes and not two of them were the same. A Japanese scientist Ukichiro Nakaya at Hokkaido University in Sapporo investigated the conditions of snowflake formation. There is now a huge snowflake institute in Hokkaido and the only place as far as I know where there is a monument in the form of a snowflake commemorating his research. Coming back to polygons and polyhedra, there is no limitation for polygons, that is, regular polygons, whereas there are only five regular polyhedra possible.

Platonic solids.

Yes. What is a regular polyhedron? It has regular and equal polygons for its faces and its vertices are all surrounded alike. They are the tetrahedron with four faces, hexahedron (the cube) with six faces, octahedron with eight faces, dodecahedron with twelve faces, and icosahedron with 20 faces. These five and no more.

These five have played a tremendous role in intellectual history, going back at least to the 700s B.C.E. from the ancient Greeks to Leonardo to Kepler to modern times. The recent discovery of carbon-60, buckminsterfullerene, is related to icosahedral symmetry. We are coming to the end of the program and have only a few minutes left. I want to skip this interesting area of the fullerenes and would like to ask you, how did you come to write this book. It is filled with pictures that you and your wife have taken in your travels. What is the background of how you came to write the book? Was it thought out ahead of time or was it some accident associated with your travels? How did it take place?

Of course, I must simplify my answer. It came from two sources. One was when our children were small, we took photographs of them, and of everything else, once you have your camera with you. Only later did we realize that symmetry dominated our images. Unknowingly, we took pictures of symmetrical objects. Later, we returned to some of the sites to find more images, but we found nothing because, apparently, we had already photographed everything of symmetry interest. So, we had been interested in symmetry even before we realized this interest. The other source was when I was preparing for my doctoral exams, I rediscovered — so this was not an original idea — that in chemistry, spectroscopy, and crystallography, symmetry is so basic and so helpful. If I think in terms of symmetry, I can spend one tenth of the time to learn about things in these fields. It helps me to classify, to categorize, to recognize patterns, and so on. It's a very important technique in scientific research as well.

I agree. We've run out of time. Symmetry is such a fundamental concept underlying physics, chemistry, biology, and all things that we interact with as we move around in life. I must end by thanking you for discussing the book.

4

ANDREA FERENCZI, "STOCKHOLM BEFORE AND AFTER" (EXCERPTS) *ÉLET ÉS TUDOMÁNY*, 2002/50, 1577–1579 (POPULAR SCIENCE WEEKLY)

This interview was recorded following the publication of my book *The Road to Stockholm*. Used with permission from Andrea Ferenczi.

A year ago [in 2001], you gave an invited lecture in Stockholm at the Royal Swedish Academy of Sciences during the Nobel Prize centenary celebrations.

In my presentation, as in my book, *The Road to Stockholm*, I highlighted not only the importance of the prize but also the fact that, although no more than nine awardees out of thousands of possible candidates are selected each year — if we stick to the Nobel Prizes in science — everyone can feel a little bit like he or she has a chance. In accordance with the founder's will, the Nobel Prize is not necessarily awarded to the greatest scientists, but for the most significant discoveries. That's why I chose the biblical quote for the title of my talk: "Many are called, but few are chosen."

Magdolna and Istvan Hargittai in Stockholm, December 2001, during the Nobel Prize Centennial, by unknown photographer.

It's a pity that no Nobel Prize has been established in the category of scientific visionaries. There are scientists whose names are not primarily associated with a discovery, but whose entire body of work has had a tremendous impact on the development of science. Leo Szilard, J. Desmond Bernal, and George Gamow certainly belong to this category, and none of them has been awarded a Nobel Prize. The name Gamow, who was a good friend of Edward Teller, is best known for his excellent books on science, which still sell in large numbers. However, he was a unique figure in both astrophysics and biomedical science. He was a pioneer of the Big Bang theory of the birth of the Universe. Then, he came up with the idea about the genetic code shortly after Watson and Crick's suggestion of the double-helix structure of DNA.

Can the Nobel Prizes be used to predict the direction of scientific progress in the future, or at least in the near future?

Nobel Prizes should not even be used to write the history of science, and they should certainly not be used to predict future scientific progress. If we look at the Nobel Prizes in chemistry awarded in recent years, we see that they were awarded primarily for discoveries that had already had a broad impact either on science or on our lives, or both. The prize looks backward, not forward: it often goes back decades to select those that deserve it.

Can we predict which field of science will develop most dynamically?

Only in broad terms. Biomedical science, which aims to improve the quality of life, is certainly one of them. In the field of physics, the most exciting scientific discoveries will be in the direction of communication and information technology. Whereas physics and chemistry used to be linked, and then chemistry and biology, today physics and biology overlap in many respects. As for predictions, the most important discoveries are guaranteed not to be the ones we would be able to predict today. Not only will there be discoveries that we now think impossible but also discoveries even whose impossibility could not be imagined today.

Is there life after the Nobel Prize? This is the title of one of the chapters in your book.

We know of a scientist who, every October, awaits the announcement of the Nobel Prize with great excitement and, as he is not among the recipients, the next day he shows up at work dressed in black. George Klein has dubbed this disease "Nobelitis." He has also defined another disease, "Nobelomania," the disease of Swedish scientists who live their lives in a frenzy about whom to give the next Nobel Prize. The Nobel laureate physicist Leon M. Lederman told me that journalists even ask him about the length of women's skirts for the next season. On the other hand, Frederick C. Robbins, a pediatrician who was part of the team that received the award "for their discovery of the ability of poliomyelitis viruses to grow in cultures of various types of tissue" and made possible the vaccination of children against polio, was never once called to consult on a sick child from the day he was awarded the Nobel Prize in 1954. He had suddenly become unapproachable, to his great chagrin. Now at 85, he is still active but has not worked as a doctor for nearly fifty years. The son of the Nobel laureate physicist Robert Wilson was himself very talented in physics, but the day his father's Nobel Prize was announced, he changed his major at university.

Evolutionary psychologists believe that symmetry plays a much greater role in our lives than we think. It reflects on health and is related to success; we tend to think of symmetrical things as beautiful – without always realizing this. What do you think about such notions?

There was a research project at the University of New Mexico aiming at uncovering the extent to which perfect facial symmetry could be correlated with the sexual lives of college girls. In my opinion, however, reducing human relationships to such geometric criteria is nonsensical and even an abuse of the concept of symmetry. The statement in the question must be countered by the fundamental fact that life is always associated with some kind of symmetry deficit: all the molecules that govern the life processes are dis-symmetrical, lacking some element of symmetry, in particular, mirror plane. Perfect symmetry is a characteristic of death rather than of life. Some consider crystals, those highly symmetrical structures, to be molecular graveyards, claiming that crystallization is death itself.

Phenomena become observable when they lack some symmetry. Pierre Curie said more than a century ago, "dis-symmetry creates the phenomenon" and this is still valid today. Crystals, although they are indeed highly symmetrical as a whole, are also characterized by broken symmetry at the atomic scales. Absolute symmetry is perhaps only achieved in total emptiness and perfect chaos.

Human perception is characterized by the fact that it completes certain gaps. If we see a damaged, imperfect cube, we still perceive it as a cube, even if the deviation from regularity irritates our senses. Each kind of observation must be evaluated with consideration of its role. We have the ability to overlook imperfections. However, if we are looking for an effect, we should pay attention even to all small deviations, because they may carry the information we are looking for.

As I understand it, the absence of symmetry is more important in life than symmetry itself.

Let me give you an example from molecular biology. The DNA molecule has a twofold rotational symmetry. The axis of this rotational symmetry is perpendicular to the axis of the DNA molecule, and the molecule has no mirror plane. The two chains run opposite each other, complementing each other, and complementarity is achieved. The same complementarity is crucial for the structure of molecular crystals. If a crystal is constructed by irregularly shaped but identical molecules, then in the most advantageous spatial arrangement there will be no mirror plane between adjacent molecules. The protrusion of one molecule would complement the indentation of the other. Such an arrangement would minimize the empty space in the crystal, so it would be the most economical arrangement. In the presence of mirror symmetry protrusion would be opposite protrusion and indentation opposite indentation, resulting in a great deal of empty space. No wonder that mirror symmetry is not found in the most commonly occurring crystal structures. I stress that I am talking about the internal structure and not the external appearance of crystals. Nature does not like unused space and can achieve this better with lower symmetry than with higher symmetry.

How do you define symmetry?

It's a difficult question. There is no universally accepted definition. However, if you ask anyone on the street if they know what symmetry is, everyone will say yes. It is a lucky concept because its meaning is consistent in everyday life and in science. An object, for example, is symmetrical if, after I have done something to it (like rotate it or shift it), I see it as it was before. My own definition is that a pattern is symmetrical if it can be formed by a simple rule. This definition applies to all symmetries and is the broadest possible definition of symmetry. It can be applied to the symmetry of a face as well as to the symmetry of a fence, which is created by the repetition of its elements, and it can also be applied to the world of fractal dimensions. The weakness of my definition is that the notion of *simplicity* is undefined, and ultimately it is up to our own judgment as to what we consider symmetrical. However, this statement is much less vague while still sufficiently flexible to cover a broad appearance of symmetry. We can refine our criteria according to the goals of our investigation.

You are preparing another book for publication.

It will be about my encounters with great scientists, their work, and their lives, as well as extracts from my own life and the lives of some of my friends. Nineteen of the chapters are titled after Nobel Prize winners, but this is symbolic. The aim is in part to disseminate scientific knowledge, and, much more. I deal with the issue of anti-Semitism, I write about our deportation — we were on a train to Auschwitz, which was diverted to Austria en route — about the *lager* [concentration camp] in Vienna, the hurdles I faced in getting into high school and then university.

The Nobel Committee for Chemistry has been asking you for years to submit nominations for the prize. Were you surprised to see the list of those who have been awarded this prestigious prize this year [2002] and not only in chemistry?

Almost all the prizes this year had something to do with correcting prior omissions though, perhaps, not in chemistry. One of the physics laureates, Raymond Davis, Jr, at 88 years of age, is one of the oldest awardees ever. He initiated and achieved the detection of neutrinos from the Sun on Earth many decades ago. One of the physiology or medicine laureates, Sydney Brenner, has been a pioneer of molecular biology and has made several discoveries for which he could have received the award a long time ago. The prize-winning novel by Imre Kertész, the laureate in literature, was first published decades ago and almost forgotten, and now the Nobel Prize lifted it out of oblivion.

5

ANDREA FERENCZI, "ENCOUNTERS, LIVES, TRAINS ..." (EXCERPTS) *HETEK*, 2003, JUNE 13, P. 14 (CULTURAL WEEKLY)

Used with permission from Andrea Ferenczi.

This is your first book in Hungarian for twenty years, and you have written more than a dozen books in the meantime. As a cosmopolitan intellectual, do you prefer to think in terms of the cultural dimension of major languages?

Ten years ago, I started to write down everything that happened to me as a scholar and as a Jewish Hungarian, thinking that if things were different in the next millennium, my children and future grandchildren should know about their roots. Unfortunately, about my parents and grandparents, hardly any written documents of our family survive, except for a few photographs. I first wrote this book in English, perhaps because I felt that all that I wanted to say, all that I wanted to face — because I chose to face it — was of no interest to anyone in Hungary, and there was no need for it. But Imre Kertész's Nobel Prize in Literature shook me up and I changed my mind, hence the Hungarian version of *Our Lives*.

This memorial plaque was erected in 2006 on the wall of the school building at 10 Bischoffgasse, Vienna, following the publication of *Our Lives* (photograph by Istvan Hargittai). Its German text in translation: "Around 500 Hungarian Jews, including children, were interned in this school in 1944–1945. Many of them died because of the deprivation and abuse suffered at the hands of the National Socialists."

You mention Imre Kertész, so let me quote him, "What kind of Jew is anyone who has not received a religious education, does not speak Hebrew, knows little of the source works of Jewish culture, and lives in Europe rather than in Israel? Someone for whom only Auschwitz means the Jewish identity, cannot be called Jewish in a certain sense."

I did not have much interest in issues related to Judaism. I grew up with a dual identity, like many people in Hungary. In 1944, when I was less than three years old, our family was deported (by then, my father had been murdered in slave labor). This was what the war meant for us. That this was because of our Jewish origin only became significant later: in many respects, our lives were the lives of assimilated Jews, with all their ambiguities. I would not venture to define who is a Jew, because it is everyone's right to decide for themselves why they do or do not consider themselves one. Everyone can have multiple identities, and most people do, whether they admit it or not. In the past decades, I had the opportunity to stay away from Hungary, but I always came back. I consider myself a Jewish Hungarian rather than a Hungarian Jew.

You were just one year old when you lost your lawyer father serving on the Eastern Front. Do you have any memories of him?

My father, Dr. Jenő Wilhelm (1901–1942), in the slave labor camp, by unknown photographer.

When I was a child, there was only one photograph in my room: my father's last photograph and he was in military uniform. He was handsome, his clothes looked sporty and smart. To me, he was the epitome of courage. I drew a lot of strength from his photograph. As an adult, I decided to enlarge this photograph. I was an amateur photographer, and my equipment only allowed me to produce small pictures. So, I did the enlargement in small sections and planned to put together the complete image like a mosaic. As I was developing the enlarged details, it dawned on me that my father's clothes were made up of worn, mismatched pieces that were not noticeable in the old photo. I was confronted with what I already knew that my father was not an elegant soldier but a prisoner. He spent the last days of his life picking up mines with his bare hands. On 30 September 1942, a mine ripped through his lower body, and he bled to death. I never put the enlarged details together.

In your book, episodes from your own life are interspersed with stories of the lives, struggles, and discoveries of Nobel-laureate scientists. How did you encounter them and why did you choose them and their names as the title of each chapter?

Over the last ten years, I have interviewed more than seventy Nobel laureates and many other scientists of the same caliber. I have published these conversations in international journals and in interview volumes. Above all, I have been concerned with the human aspects of these researchers and their discoveries. I have often heard that the results of scientific research are independent of the personalities of the researchers. In many cases, my meetings and conversations disproved this. It was also my intention to bring scientific questions and discoveries closer to the reader through a personal approach. The reader of my interviews meets physicists, chemists, and biomedical scientists. The more I investigated the lives of these people, the more I discovered common features in their lives and careers.

Even in cases like the German Nobel laureate Manfred Eigen, and the Budapest chemistry professor László Kiss, who was the only survivor of his family of five, and came home from Auschwitz?

After I finished the manuscript of the book, I sent each Nobel laureate the chapter they symbolized. Manfred Eigen (1927–2019) didn't respond for months, then one day he called me, indignantly, and said he had nothing to do with Auschwitz. We talked, and eventually, he understood that the parallel between the two lives was not directed against him, that their fates had not been linked by me, but by history. I had given it a great deal of thought to whether it was right to tell the story of László Kiss in the chapter symbolized by Eigen. They were both children at the time of World War II and by no means responsible for what had happened to them. However, I did see a connection between their fates. Eigen was training to be a pianist but was unable to practice during the war years. Their home was bombed, his brother was killed, his piano was destroyed, and his artistic ambitions were dashed. He still gives concerts with the Boston New Orchestra and the Basel Chamber Orchestra, in addition to his academic work. Kiss lost his entire family, and he and his twin brother knew Dr. Mengele — SS Hauptsturmführer Mengele — closely. When Kiss returned from Auschwitz to Seregélyes, his hometown, he found only the family piano. It was in their neighbor's yard, being used for storing hay. Eigen became a great scientist, recognized worldwide for his achievements. Kiss retired in 1996 from his position as head of department at Eötvös University. His greatness as a human being is reflected in his ability to build a new family around him after all he had been through and to live a full life in an environment that pretended Auschwitz had never existed, Manfred Eigen was awarded the Nobel Prize; László Kiss's prize is the new life he created.[1]

In addition to interesting facts about the history of science, your book deals with science policy issues. It is surprising that the perception of the conservative political physicist Edward Teller is so different in the United States and in Hungary.

Most of the science community in the United States has a negative opinion of Edward Teller. There are several reasons for his unpopularity. During the Second World War, while the first atomic bomb was still being worked on, he was impatient to get the hydrogen bomb made. He also made enemies by pushing through the opening of a second weapons research laboratory in California, in addition to Los Alamos. But the most severe damage he did to his

---

[1] László Kiss (1928–2022) and his twin brother were subjects of the infamous Josef Mengele's human experiments in Auschwitz. Immediately after his return, Kiss put his ordeal on paper. I quoted extensively from his writings and our conversations in my book *Our Lives.*

standing among his peers was the way he spoke at the security hearing of his physicist colleague Robert Oppenheimer in 1954. While he expressed confidence in Oppenheimer's loyalty, he was unambiguous in having no trust in his handling of crucial defense matters and thus would deny Oppenheimer security clearance. After Teller's testimony, Oppenheimer's access to classified work was terminated. Although Teller's views were shared by several of his colleagues, they were more discreet. He was also the main proponent of President Reagan's so-called Star Wars strategy. He should not be blamed for this as the Star Wars strategy probably contributed to the collapse of the Soviet Union, even if only indirectly, because it forced the Soviets into an arms race that they could ill afford. Teller's impact on science policy and U.S. military policy is likely to be the subject of intense debate for a long time to come.

Not so at home, because last year [2002] on March 15 in Kossuth Square, one of your academician colleagues read out Teller's statement on the elections supporting, even if indirectly, Viktor Orban and his party.

Edward Teller was then already ninety-four years old. I don't want to underestimate his intellectual capacity due to his age, but it is unlikely that he could have judged what his statement would be used for, nor is it possible to know how authentic that statement was. In 1996, when my wife and I visited him at his home in Stanford and recorded a conversation with him, the political situation in Hungary came up in our conversation. Surprisingly, he praised Gyula Horn, the leader of the party that is the successor of the former communists. He said that he appreciated Horn's ability to open a modern direction preparing for significant changes. This is also Edward Teller, who is universally regarded as a staunch anti-communist. However, I do not think it is alright to use Edward Teller in any context in today's Hungarian domestic politics.

Why do you think societies in Central and Eastern Europe were more susceptible to Nazi ideology than Western democracies?

I would not make such a strong distinction. In a democracy it is not a question of whether we like Jews or not, but of how we treat our citizens. I do not know whether Jews in Denmark were loved or not, but they could only imagine one kind of behavior towards their fellow countrymen: solidarity. Furthermore, I would not distinguish between Western and Central and Eastern European attitudes so sharply, because we know of very sad stories in France, too, but at the same time, something like what happened in Denmark happened in Bulgaria.

In your opinion, will there ever be an in-depth examination of whether and how much responsibility, if any, was borne by certain strata of contemporary Hungarian society for the massacre of close to 600,000 Hungarian Jews?

A few years ago, I read an article in which the author hoped that there would be no Jews to sue young people shouting Nazi slogans at football matches. The slogans were that the trains would start again with the remaining Jews — to Auschwitz. In my opinion, this is not a Jewish problem, but a Hungarian problem. The metaphorical departure of the trains to Auschwitz in 2000 and 2001, and the silence outside the stadiums as a backdrop, were evidence of something. Something that included nostalgia for the semi-fascist regime of Nicholas Horthy with his re-burial, and the statue of Pál Teleki, who in 1939 presented the second Jewish law to the Hungarian Parliament with the claim that if it had been up to him, he would have made it harsher. It also included the fact that in 2001, Hungarian public television showed a film about Ferenc Szálasi, the leader of the ruthless Hungarian Nazi movement, the Arrow Cross. The film was trying to prove that Szálasi was not an anti-Semite. In light of the fact that during his reign, conservative estimates suggest that at least a hundred thousand Jews were murdered in Budapest, and that women, old people, and children were thrown into the Danube after being tortured, it is irrelevant whether Szálasi was anti-Semitic or not. What is to stop a young fan who sees Horthy's reburial, walks past Teleki's statue, and watches a public TV program about Szálasi, from shouting "Sieg Heil" at soccer matches and be willing to send the trains to Auschwitz again? Today's anti-Semitism in Hungary is more criminal than before. Whereas before 1945, Auschwitz was not known, today all manifestations of anti-Semitism are made in the knowledge of the Hungarian Holocaust. I had hoped that Imre Kertész's Nobel Prize in Literature would confront many people with this, but the mixed reception his award has been met with does not bear this out. This is not something that could be viewed from a sterile professional and aesthetic point of view and reduced to a literary work, even if Kertesz's prize-winning book *Fatelessness* becomes compulsory reading because we all know what fate compulsory readings have.

6

ISTVÁN PALUGYAI, "THE HOBBY OF AN ACADEMICIAN: DOING INTERVIEWS" (EXCERPTS) *NÉPSZABADSÁG* (FORMER DAILY NEWSPAPER), 2003

Used with permission from István Palugyai.

The first questions were about the interviews with famous scientists.

In my interviews with famous scientists, I always stressed that I was not a journalist, but a colleague. The interviews often went into serious technical depth, which is what all scientists like, and they could not help revealing also details of the human side of their lives. One of my subjects stressed that he wanted to stick to the technical facts. However, when I sent him the transcript, he wrote that, to his surprise, he had told me things he had never shared even with his wife. Moreover, he did not delete anything from the text. Of course, each interview is different, none of them follows a standard model. At one extreme, the interview with a Nobel laureate in the countryside, sitting in a car, was completely spontaneous, while at the other, a

Japanese Nobel laureate asked me to send him the questions in advance. I received his answers and when I visited him in Kyoto, I asked more questions. He wrote them down and a week later I had his answers. Of course, I can see now that I was very lucky at the beginning that none of my perspective interviewees slammed the door in my face.

Selection? Preferences?

The occasions are random. Usually, a conference or a study trip gives me the opportunity to meet a great mathematician, physicist, chemist, or biomedical scientist, or when one of them comes to Budapest. Meeting them in Budapest is the hardest because their contacts, the ones who had invited them, are often jealous of whom they get to meet. Over time, it has become easier to find interviewees. Once I was a Visiting Professor at the California Institute of Technology as invited by Ahmed Zewail (he was 2 years away from receiving his Nobel Prize). Ahmed introduced me to a recent Nobel Prize winner [Edward B. Lewis] and over dinner it became clear that he would be happy to be interviewed. It came off and went well. In this case, it was interesting that this Nobel laureate had previously stated that he was reluctant to be interviewed. When his Nobel Prize was announced, he was away from his university and had glued a chin beard on his way home so that journalists would not recognize him. This was, of course, as much a pose as a genuine aversion to publicity.

Was there any unsuccessful encounter?

A few years ago, I was in Cambridge, England, talking to a famous scientist. When I took down the transcript, I decided not to use it because it was full of platitudes. I wrote him about my problem honestly and the next time I was in Cambridge we repeated the interview. Unfortunately, with the same result. It started well, but after only a few sentences it slipped into what was obviously a habitual text that had developed over the years, and which was uninteresting.

There was also, at the beginning, a test of how much I wanted the interview. On one occasion, a conference in Los Angeles gave me the opportunity to ask a famous chemist for an interview. He gave me an appointment, not nearby, but at his vacation home several hours' drive from the city. I hired a car and drove there, surviving two detours, and the interview went brilliantly. Years later, he told me that when he got my invitation for the interview, he went to the library to learn about me and decided to test me, hence the faraway location for our meeting.

There have also been times when technology has let me down. I went to see the Nobel laureate medicinal physicist Rosalyn Yalow at her workplace, the Brooklyn War Veterans Hospital. I was shocked to find her in a wheelchair, with a stroke. This showed that I had not prepared enough, because I should have known about her condition. Had I known what had happened to her, I might not have asked for the meeting, which would have been a shame. Although speaking, uttering every single word, meant a tremendous effort, she was happy to talk about herself and her work. It was probably also due to my awkwardness in this situation that I only realized at the end of the conversation that my tape recorder was not working. It was a terrible feeling, but there was no way to start again. When I said goodbye, I sat out in the hallway and said everything I could remember into the tape recorder. When I sent her the "transcript" of our conversation, she corrected just a few minor details, and I breathed a sigh of relief.

7

SZILVIA KRIZSÓ, *INTERPRESS MAGAZIN*, JUNE 2004, 48–51 (EXCERPTS)

Used with permission from *Interpress Magazine* and Szilvia Krizsó.

It is not usual for a scientist to interview other scientists.

This had two origins. Back in 1965, at the request of Radio Budapest, I interviewed the Nobel laureate Soviet scientist Nikolai Semenov. I went unprepared, I was almost arrogant in my behavior, but the interview went well. Of course, this was to Semenov's credit, as he was an experienced interviewee. He also answered questions that other people don't, such as where he thought science would be in 30 years' time. He made predictions and, listening to the interview 30 years later, he was not far wrong. In the late 1990s, I spent six years editing a New York-based science magazine for chemists [*The Chemical Intelligencer*] that I founded, and I started interviewing scientists for it. These have been so well received that a London publisher has already published four volumes of 36 interviews in each with two more planned [*Candid Science*].

From the interviews and from the background research for your book on the Nobel Prize, did you form a system in your mind of when, in what field, and who gets the next one?

There is no system, and the Swedes deny that there is any. Perhaps coincidences could be discovered, but — because of the small number of prizes — this cannot be demonstrated. It is therefore impossible to predict who will receive the next one. We can only talk about having a system if it is good for prognostication.

Do you ever think that you will win this for chemists? After all, you spend so much time on the Nobel Prize.

I can deal with this subject objectively and, for me, in an entertaining way, because I know that I have no vested interest in it. But I also know how much anguish it causes for those who come close to getting the prize, which they either get or they do not. From October to October, they live under terrible pressure, even if they are truly modest and perhaps not really expecting it, but their environment does not allow them to take themselves out of that pressure. However, studying Nobel Prize-winning discoveries and the researchers involved has been a real school for me. Thanks to this, for example, I have developed a love of biology, and I am gaining a wealth of information. For example, I can tell who I think is the most likely to win the Nobel Prize in science in Hungary.

Who is it?

I stress that I cannot talk about whom I nominate when I am regularly asked to do so because I am bound by strict confidentiality. I can only tell you whom I would expect to receive it. That is Árpád Furka, a former professor of organic chemistry at Eötvös University and the initiator of a new direction, combinatorial chemistry.

Did you have a mentor?

Several. For example, I learned to write articles from reading articles by certain scientists. There are many people to whom I owe a lot. In high school where I had an adventurous path, it was my history teacher who set an example. Then I had a horrible chemistry teacher, but I learned a lot from him, so I'm very grateful to him too.

It was an adventurous journey to high school. Why?

I went to primary school in Orosháza, a small town in the southeast of Hungary. There I applied to the local high school and was accepted. But a week before school started, they canceled my application on the grounds of my "class-alien" social status. What happened was that one of my classmates was not admitted on the ground that he came from an industrial kulak family. To this day I am ashamed that we were not sufficiently outraged about this. His father appealed. He wrote that while his son had not been admitted, seven of his classmates had been admitted, even though they came from similar families. My grandfather used to be a shopkeeper in Orosháza. They examined the appeal, found it justified, and annulled the admission of the other seven children as well; I was one of them.

Mother with my brother (right) and me (left) in the garden of our home in Orosháza, by unknown photographer.

How did you cope with this after you and your family had already been deported by another political system also because of your social origin though for a different reason — it was a miracle that you survived?

Perhaps this gave my mother the strength not to let it go. She even went to the county secretary of the communist party, who said that it was not possible to prevent a "class alien" from going to university early enough. But it was 1955, and the winds were changing in Budapest. An article about our case appeared in the public education journal under the title "These are our children now." The Ministry decided to admit me to any high school in the country except for our local high school. They did not want to offend the sensibilities of the local authorities. That's how I ended up in Budapest.

It seems you collided with history at every step.

That was not how I felt about it. Even when thinking about our deportation I just thought how lucky I was to have escaped. Today I see it differently, and I am outraged that I was locked up in a concentration camp at the age of three. But it didn't make me gloomy, it just made me realize that I was only lucky in a great horror story.

What did you owe your survival to?

There was a Jewish group that negotiated with the Germans to rescue people in exchange for goods. From the Germans' point of view, this had some additional merit as some of the Nazis were already thinking about the repercussions against them after the war. Furthermore, there was a great need of labor, let alone cheap labor, in Austria, especially in Vienna. My mother was made to work as a roofer, for example. Some eighteen thousand Jews were "put on ice," and directed to Austrian labor camps. We were not among those eighteen thousand initially, but one of their trains was sent by mistake to Auschwitz, and our train, already en route to Auschwitz, was diverted and sent to Austria instead.

And how did you get back?

The Red Army liberated our camp, and we went home. My uncle pushed me on a wheelbarrow most of the way. It is interesting to me that my mother could not imagine going in any other direction. When I think back — and I must add that I am happy how my life has turned out — I would not have started on the way back. Because it was not just the deportation, it was the end of a process. And that's why it bothers me that today people are trying to ascribe the Hungarian Holocaust exclusively to the German occupation of Hungary. This was as much a Hungarian action as a German one. And that is what we should focus on first and foremost. The Germans, for their part, have already done that, but we have not yet. In this, I believe that Hungarian society has not yet reached the level of self-awareness, or, rather, self-respect, to face its past honestly.

8

V. V. BLAGUTINA, "DEEP DRILLS AND SCRAPERS." *KHIMIYA I ZHIZN*, 2006, NO. 7, 20–21 (RUSSIAN POPULAR CHEMISTRY MAGAZINE)

Do you have anything in common with the scientists you talked to? What does it take to be a successful researcher?

These great minds are mortal people just like you and me. In many cases, in more cases than I would have originally thought, they have had a difficult life: their childhood and youth were far from carefree, and they did not always start their careers in favorable circumstances. Many of them attributed their good fortune to chance and saw themselves as just being in the right place at the right time.

I see two important factors in the development of a scientist. The most important is the mentor, who is not necessarily a real live person, because it could even be a literary figure. Then, the influence can be the result of a long-standing relationship, or it can be a passing comment from a colleague. The environment in which a newcomer starts is important. Some research centers have produced several Nobel laureates in the past, but real talent can flourish anywhere, especially if there is a right partner to work with.

As for human qualities, I can only mention very subjective things. I think a good scientist has both democratic and autocratic traits.

Scientific issues are not decided by his own authority, but at the same time, they are not decided by voting either.

Curiosity is an indispensable characteristic; the true researcher does not shy away from the unexpected and does not design experiments whose results are those that are to be expected. Curiosity and a receptive mind help to develop versatility and make the researcher feel at home, at least to some extent, in many fields. A broad basic education helps to orient oneself in completely unknown territories when one may be close to discovering something new. Often, the scientist does not realize the significance of the discovery and may only recognize it in its fullness years later, when others have successfully applied what the scientist had first discovered.

True greatness is not a pursuit of glory, but a desire to understand something. It would be a pity, however, to consider it degrading to try to make the discoverer of the discovery to be known. Failing this, the discovery itself may disappear into obscurity.

There is no single type of scientist, and diversity is also a feature here. It is an oversimplification, but an interesting classification divides scientists into two groups. There are deep-drilling types and there are scraping types. The deep digger tries to learn as much as possible about a narrow field and may devote an entire career to a single topic. The scraper moves from topic to topic. But it is impossible to predict which one will be more successful. Both types are found among the greatest.

Can we talk about discoveries that are ahead of their time and discoveries that are ripe for the taking, and if someone doesn't do it, someone else will sooner or later?

Of course, it is not that simple. It makes no sense to say, for example, that I am making a discovery now, and it makes no sense to say in advance that I am going to make a discovery before its time or one for which the time is ripe. In the case of a premature discovery, much of the value of the discovery is detracted if it is very much ahead of its time, not to mention the difficulty of distinguishing the valuable from the worthless discovery if it is indeed too early. If the discovery cannot be applied, cannot be exploited because it is not yet in demand and the community is not receptive, it may indicate that it is ahead of its time. Gregor Mendel's genetic discoveries remained in obscurity for decades and it is fortunate that by the time they became timely, they had not been completely forgotten. The discovery of the C_{60} buckminsterfullerene molecule was not only timely but could even be considered a belated discovery, as was Aleksander Nesmeyanov's idea of atoms or groups of atoms encased in a carbon cage.

It is often said that if one scientist had not made a discovery, sooner or later another would have. This can be contrasted with works of art, because if Rembrandt had not painted one of his paintings, surely no one else would ever have painted it. But scientific discoveries also have their own individual style. If Watson and Crick hadn't discovered the double helix structure of DNA, someone else would have, and probably not very long afterward. But it was only Watson and Crick's discovery that characterized their hypothesis, which was originally a suggestion rather than a real discovery, in a sudden and single stroke. Others would have arrived at the double helix structure, but probably more slowly and in a series of small steps rather than in one fell swoop, and the effect would have been slower and more gradual.

The Nobel Prize is the most prestigious scientific award, but can we write the history of science based on the Nobel Prizes?

Such a history of science would be distorted and misleading. There is no doubt that Nobel laureates are excellent scientists, but not always the greatest. In his will, Alfred Nobel was very clear: he wanted to reward the most important discoveries, regardless of how great a scientist the discoverer was. Then, the Nobel Prize was not awarded for every important discovery and the awarding of prizes was often followed by protests and discontent.

Furthermore, science is not just about discovery. It is essential to work in detail, to organize knowledge. According to my former thesis reviewer and translator of some of my books into Russian, the internationally renowned crystal chemist Petr Zorky, the role of a scientific explorer is like that of a military scout. The heroic explorers are fol-lowed by the front line, which is led by excellent commanders. For science to follow on from these discoveries, it needs the work of many other scientists. Once the domain of a select few, science is now almost industrial in scale. Of course, there is still a role for special minds to make discoveries and set the main directions of progress, while many researchers carry out routine tasks that are less exciting but just as necessary as the truly front-line research.

In which country in the world do you think science is most important?

I would divide the answer into two parts. Which country gives the most to science and which country has the greatest respect for science? The United States devotes the most to science and probably the former Soviet Union had the highest respect for science. The former is much more effective than the latter.

9

TECHNION TELEVISION, HAIFA, 2011, JANUARY 12 (INTERVIEW WITH PROF. ISTVAN HARGITTAI (YOUTUBE.COM), EXCERPTS)

There was a question about the initial interest in Dan Shechtman's discovery of quasicrystals.

I have been doing research in molecular structures, have been interested in symmetry, and have been fascinated by the discoveries in 20th-century science. All this combined for me to become interested in Danny Shechtman's discovery of quasicrystals. The discovery happened in 1982 and was published in 1984.

What is the essence of his discovery?

He discovered condensed-state structures that have fivefold symmetry which used to be considered impossible. And it was not only his discovery but also how he carried through the years while many people including very famous scientists did not believe that it was a genuine discovery for which he deserves a lot of credit. Maimonides is attributed of having said that just because you read it in a book, it may not necessarily be true. This is exactly what we could see in Danny's work. He observed something, which was contrary to everything, all dogmas, but he went ahead and showed its validity and eventually, he convinced the world of science.

How did your interest in the subject of discoveries happen?

Some years ago, I developed a new course at the Budapest University of Technology and Economics about great discoveries in 20th-century science. First, I started with very few students and gradually it developed into a popular course with about 60 or 70 students. This was also why it was losing its appeal to me because it was no longer an intimate discussion forum; rather, a series of formal lectures. Shechtman's example is more than an interesting episode in science history. It provides an opportunity to discuss the nature of scientific discovery of how you make a discovery, and what are the merits and demerits of the discoverer. It does not suffice just to make the discovery; you must prove it to the world that it is indeed something new. Sometimes the discoverer himself/herself may not recognize the importance of the discovery or that there was a discovery in the first place. Danny did recognize that he made a discovery, but I doubt that he recognized at once its full implications.

Over the years I conducted a large project of interviewing scientists. My wife and son joined me in this project, and we produced six volumes of in-depth interviews. Each volume contained 36 interviews, at least half of them were with Nobel laureates. I have also written a book about the Nobel Prize, *The Road to Stockholm*. One of its features was the difficulties many of the future laureates had to overcome on their way to top science. Another feature was the comparison of pairs of scientists of comparable performance and achievements and one of them becomes a laureate and the other does not. Some people are prone to recognition, others, not so much. What personal traits characterize those discoverers that get plenty of recognition while others may not? Scientists are also human beings who crave recognition although many deny this. And, of course, there was a large chapter of those scientists and discoveries that should have received the Nobel Prize but did not.

Danny Shechtman was one of my interviewees and his discovery is an excellent example, the kind of achievement for which the Nobel

Prize was created. It is given for specific discoveries rather than for lifetime achievements. Shechtman's discovery is well-defined and easily delineated from other peoples' contributions. There could be no controversy in awarding him the Prize.

The Hargittais (left) and the Shechtmans (right) at a reception at the Royal Swedish Academy of Sciences, December 2011, Stockholm, during the Nobel Prize award week, by an unknown photographer.

[Note added: Danny Shechtman's Nobel Prize was announced nine months later, in October 2011, for his quasicrystal discovery. The interview with Shechtman in the *Candid Science* series is one of about a dozen interviews with scientists who were not Nobel laureates at the time of the conversation but would receive the Prize later. The "delay" could be two months (Avram Hershko) to 20 years (Roger Penrose).]

10

JÓZSEF HUBERT, "DETERRENCE IS THE KEY TO PEACE" (EXCERPTS) *HETEK*, 2011, JUNE 24, 20–21 (CULTURAL WEEKLY)
Used with permission from *Hetek*.

*I*s *Edward Teller a scientist who deserved the Nobel Prize but didn't get it, or is he as a nuclear physicist responsible for the deaths of masses of people?*

Let's start with the two atomic bombs dropped on Japan, which did indeed kill tens of thousands of people, but the swift end to the Second World War saved the lives of hundreds of thousands of American and allied soldiers and millions of Japanese. To return to the first part of the question, Edward Teller could have been awarded the Nobel Prize just as easily as he was not, but his missing Nobel is not the measure of his achievements.

What is it measured by?

With effects that go far beyond the Nobel Prize. Several times, for example, the Nobel Prize has recognized the discovery of an elementary particle that was important for understanding the world, while Edward Teller's work has influenced the way the 20th century has been shaped. In this case, the Nobel Prize is not a major consideration, although, I might add, Teller's impact on world events is sometimes overestimated. Many people believe that Teller imposed his will on the U.S. government when President Truman decided in 1950 that the United States would develop the hydrogen bomb, or when President Reagan announced the Strategic Defense Initiative in 1983. But this is an exaggeration, even if he himself has occasionally fueled these exaggerations. His role was to lobby outside the scientific world — in the circles of military and civilian leaders. In the late 1940s and early 1950s, few American scientists argued for the development of the hydrogen bomb, most opposed it, whereas we now know that it was essential. The Soviet Union had already decided to develop thermonuclear weapons [i.e., the hydrogen bomb] in 1946, before it had an atomic bomb, while the American scientists were still engaged in endless debates.

So, the real reason for the development of the hydrogen bomb was the existence of a dictatorship that would sooner or later acquire a thermonuclear weapon. The question rightly arises whether democracies therefore must fight offensive dictatorships.

Let me split the question in two. If we are examining whether democracies should be strong enough to defend themselves against dictatorships, then my answer is a resounding "yes." If they are not strong, they can easily cease to be independent, autonomous democracies. See the case of Switzerland, which has been neutral for centuries but has a very strong army and defense policy.

The second question is more complex: at what level can democracies fight against dictatorships? In 1956, it was a vain expectation that the Western democracies would intervene in the Hungarian revolution. In this struggle, there are severe limits to the possibilities of democracies.

Can the self-defense potential of democracies be measured in terms of the number of thermonuclear weapons or other bombs?

Only democracies with strong defense capabilities can protect themselves; dictatorships understand no other language. In the Cold War, neither side wanted to attack the other because each knew that the blowback would be lethal. Scientists have known since the late 1940s that the more terrible the weapons, the less likely one power is to attack the other. That is why eminent scientists like Leo Szilard and others have made it a priority to create ever more destructive weapons. It is a terrible paradox, but only with such weapons could they hope to maintain peace. Szilard caused enormous consternation when he proposed that nuclear weapons should be coated with radioactive cobalt to wreak even more terrible destruction. Yet he did so precisely to make the idea of war as much a deterrent as possible so that no one would think of starting a nuclear war.

Is there a difference between a scientist making such a proposal in a democracy and one in a dictatorship?

Of course, there is, and we have examples of both. We have already seen the former with Szilard, and the latter is the example of Andrei Sakharov, who later won our respect as a human rights campaigner. When he was still supporting Stalin's regime, he proposed sending a submarine to a major American port city with a hydrogen bomb and detonating it. Notice the difference: this was not a proposal for deterrence, but for implementation.

What responsibility then does the scientist have to use the discovery?

Many people are still of the opinion that the scientist's responsibility ends when the discovery is made because it is up to society to decide how to use it. However, the researcher also has a responsibility to inform society about the possible consequences of using the invention or discovery. Coming back to the example of Leo Szilard, he went even further: he launched a movement for a more livable world by advocating the election of senators in smaller American states who identified with the policy of peaceful coexistence.

How many people were involved in the decision to use the atomic bomb?

Ultimately, one, President Truman, after listening to a small number of advisors, including some leading scientists.

So, the public learned about it after it had already been dropped over Japan.

That's right. There is a palpable contradiction between a democratic, open society and a secret project. Two billion dollars were spent at the time on the Manhattan Project to develop the atomic bomb, but nobody could have known about it. Truman consulted a four-member scientific committee before acting. They also advised on where to drop the atomic bomb. The Americans deliberately left some Japanese cities untouched during the war — including Hiroshima and Nagasaki — so that they could measure the effects of the atomic bomb. I point out the little-known fact that carpet bombings of Tokyo caused more casualties than the atomic bombs.

However, the deterrent effect was much smaller.

In addition to forcing Japan's surrender, the strategic payoff of the atomic bomb was to deter the Soviet Union, which was increasingly seen as the next adversary of the United States. But we also know that attacking the Japanese islands would have resulted in terrible casualties on both sides. The Japanese were desperate, they regarded the emperor as a deity, and they would have fought with extreme blindness. After the first atomic bomb, nothing had changed, their war cabinet wanted to continue the war, and it seems that a second one was needed to make the Japanese leadership and the emperor understand that the war was over.

Teller never moved back to Hungary, but he did return a few times, never denying his Hungarian origins and roots. If Teller had stayed at home and survived the war, do you think he would have achieved the same success as he did in America?

It is highly unlikely that he would have survived the persecution of the Jews, and the Teller family was deported and persecuted by the next regime in the 1950s. Nor, I believe, would he have achieved his scientific discoveries, which of course is only conjecture, but there are examples to draw from. Look at the organic chemist Árpád Furka, who discovered a new direction in chemistry, combinatorial chemistry. He stayed at home and received little support and recognition. His belated recognition has meant little in terms of his scientific advancement. Hungary is struggling to create the conditions in which these talents can flourish. The Academy of Sciences now has a program to bring home expatriate talents who have felt that the only place to develop them was in the West.

Is it just a question of money?

No, envy and jealousy are part of the problem. There are so many components to why it seems that young talent is not getting a chance here. In many cases, older researchers "suppress" young talent, sometimes almost forced by external circumstances, because, for example, a successful grant application might require their name to be included in a publication. This is not a problem for the government but for a truly democratic, competitive, and open society. Let me go back to the beginning of our discussion here: a democracy should demonstrate strength to the outside but should practice tolerance on the inside. Teller has beaten almost everyone in his public debates, but when, for example, his opponent ran out of time, he gave him some of his own time to finish what he had to say. This is a practical example of what Teller always said: you don't silence your opponents; you listen to them. That is not the way I see things developing in Hungary today.

When judging Edward Teller's career, to what extent can one detract from the fact that he was of Jewish origin?

Teller said of himself that he was a three hundred percent human because he was one hundred percent Jewish, one hundred percent Hungarian, and one hundred percent American. His life and career cannot be seen as independent of all this: the hopeless conditions in Hungary (which were hopeless not only for Jews) forced him to go abroad, and then the Nazi rise to power forced him to leave Europe. Even so, he did not think that he had to turn his back on his

Hungarian identity, which he not only proclaimed but also represented in his actions.

In your Teller biography, you tried to present not only the person of Edward Teller but also his immediate and wider environment. Do you see any similarities between the Budapest of his day and the Budapest of today?

Teller lived in two different worlds in Hungary: the happy peacetime he heard about from others, and the 1920s, when it was common for Jewish students to be beaten up by nationalist students at the Technical University. These two worlds were very different, yet both lived in him. Personally, I see the current restoration efforts of inter-war Hungary as a misguided path for the country. The entire Horthy era contributed to the fact that scholars such as Eugene P. Wigner, Leo Szilard, Edward Teller, and many others left the country. And let us add that this is not just a problem of a minority. It is not just Jews and people of Jewish origin, who feel uncomfortable in the atmosphere of the restored 1930s, but anyone, regardless of their origin, who wants to create rather than worry about where they come from.

I I

VERA SILBERER, "DISSEMINATION OF KNOWLEDGE AND CULTURE" (EXCERPTS) *MAGYAR KÉMIKUSOK LAPJA* (HUNGARIAN CHEMISTS' MAGAZINE), 2012, JULY–AUGUST 67(7–8), 229–232

Used with permission from Vera Silberer.

From an earlier conversation, we learned that your interviews with famous scientists were inspired by the magazine, The Chemical Intelligencer.

The Chemical Intelligencer was a quarterly magazine promoting the culture of chemistry ... I founded and edited it for six years until its publisher was sold.

What does "the culture of chemistry" mean?

It means drawing a line between two letters and writing a whole book about it. Somebody once drew the line that represents the chemical bond, and we draw it the same way today — even though we know a lot more about it than when it was drawn for the first time. It goes beyond laboratory work and requires that we explain and describe what is happening; we can somehow predict what should be done and if we do something, what will come of it.

I have looked at what The Chemical Intelligencer *had to offer its readers. Here are just a few headlines: models, mysteries, chemistry history, new applications, stamps, cooking, debates, links to other sciences and the humanities, great discoveries, and great chemists.*

Culture includes how things happened, who did them, what they did, what is worth reading about, and who to talk to about the topics we are interested in. The literature is full of success stories, whereas the exciting stuff is often born out of failure. Take the noble gas compounds, for example. The idea of these came up when the "inert gases" were discovered. Some researchers immediately tried to produce their compounds but failed. Over the decades they tried several times and never succeeded. Neil Bartlett broke through the "inert gas barrier" in 1962, and within months, not years but months afterward, many people had succeeded in making noble gas compounds independently of him. The same happened with the discovery of quasicrystals, but not in experimental but in theoretical work. Levine and Steinhardt had in fact already formulated a theory, but they did not dare to publish it because they considered it impossible to form quasicrystals. They only published their theory when they learned of Shechtman's experiment, and then hundreds and hundreds of publications followed.

The discoveries that I find most exciting are the ones that set such a process in motion. That is why I feel that the discoverers are very brave but terribly lonely people: they know something that nobody else knows and they must come out with it. Often the greatest scientists are not the greatest explorers. The greatest scientists know very well why things are not or cannot be. Shechtman couldn't explain his discovery, nor could he really describe it. But he did come up with it, a little innocently, but very stubbornly — this stubbornness being his distinguishing quality that was needed for him to succeed.

Linus Pauling was the first in a long line of extraordinary people to be interviewed in your magazine, and later in your famous Candid Science *interviews books.*

Before I started editing *The Chemical Intelligencer*, I wrote to fifty or sixty outstanding representatives of the science of chemistry and asked them if they thought such a publication was needed. Linus Pauling said it was a very good idea, but he was too busy to write for it. I didn't even think of asking him to write an article, because he was so famous and over 90. But his reply gave me the idea for the interview.

With George A. Olah (left), 1996, at the University of Southern California, by Magdolna Hargittai.

I also received important support from George A. Olah in my "opinion poll." Later, we developed a very good relationship, perhaps because I contacted him a few months before his Nobel Prize. The Nobel Prize is a watershed. A Nobel laureate will have a lot of acquaintances, but before that, he or she will not have been sur-rounded by so many people. By the time I founded the magazine, I had already had a broad circle of contacts and could ask a lot of people if they wanted to send me an article.

It seems that you prefer examining the careers of great scientists from the distance of considerable time.

I have returned several times to the history of the great Hungarian generation of physicists with emphasis on their work, and their dis-coveries. What were their roots, their sources, and how can their work be placed in the "science of the world"? Why did they leave Hungary and what would have become of them, what would have happened to them if they had stayed in Hungary? If we do not face up to this, we will not get to the point where we value our talents today, and then the exodus of researchers will continue.

My current book project has evolved as the great Hungarian physicists have led me to Soviet physicists. I speak Russian and have been interested in science history in the Soviet Union, but even for me, it was a big surprise what I uncovered. What great scientists lived there! But most of them were isolated not only linguistically but also politically, and not only because they were often forced to do secret research but also because in Western culture we traditionally pay attention to West European and North American events and currents.

Yet fantastic discoveries were made in the Soviet Union. Nikolai Semenov and a young colleague — whose name was classified for a long time because he was later the scientific director of the secret Soviet nuclear weapons laboratory, Yuliy Khariton — discovered the branched chemical chain reactions. Leo Szilard's nuclear chain reaction is also a branched chain reaction. Metaphorically speaking, the two discoveries met in the atomic bomb. Until then, a series of further discoveries were needed. But the necessary "culture" was already in place.

From January [2012] your Academy-supported research group has ceased to exist. Though this did not make anybody unemployed, this was not a welcome development.

The research group has ceased to exist with the stroke of a pen of the authorities. The saddest thing is that while we are barely able to attract back talented young expatriates from abroad, we are firing (other) talented young people at home without a word. We have just found a brilliant place in America for one of our excellent young colleagues, but we wanted to let him go with the promise of welcoming him back ... The closure of our group is not a problem for us, my wife and me, because we have formally retired, but because it has condemned a culture of enormous value to disappear. I do not know what criteria were used to make the decision, but it is a common fault of our system of research support that the criteria are revealed after the event or, as our case shows, not even then. I will not stop what I am doing, whether there is a research group or not. I cannot even get angry with anyone because of myself. But I can, for others.

12

TAMÁS UNGVÁRI, ATV TELEVISION, 2012, JUNE 9. PUBLISHED IN TAMÁS UNGVÁRI, *SZOKATLAN BESZÉLGETÉSEK* (UNUSUAL CONVERSATIONS), BUDAPEST: SCOLAR, 2013, 124–132

Would you tell me about your latest book?

It's called *Ambition and Curiosity*; it's about 15 scientific discoveries, from the explorers' point of view. I was fascinated by finding out what qualities enable scientists to discover something — which is not the same as doing science, because not all scientists are also explorers. I would also venture to say that the greatest scientists are not the greatest discoverers. Among other things, some of the greatest scientists think they know why new discoveries can no longer be made. Some "lesser" scientist, when they get their hands on a gem that is completely new, will notice it, even dare to say about it. They will not be afraid to risk their hard-earned prestige, which may not even exist yet, and may become great discoverers and Nobel laureates.

As early as the end of the 19th century, it was declared that everything that needed to be discovered and could be discovered — had already been discovered. Such voices can still be heard today. Just as

With Tamás Ungvári (left) on Vörösmarty Square, Budapest, at a book festival, by Magdolna Hargittai.

at the end of the 19th century they did not know what they were talking about, they should not be believed today either. It is not possible to foresee the future. The greatest discoveries are not the ones today we think impossible, but the ones we don't even think of as impossible. We cannot predict them. There are, of course, expected discoveries that we know will happen sooner or later. Such was the case when the Higgs particle was predicted, and we waited years for it to be found. But then there are completely unexpected discoveries that we never thought possible, or even impossible, before. Therein lies the advantage of geniuses over great scientists. A genius can see connections between different fields that others miss.

It's unpredictable who will become a great explorer, who will become a hard-working scientist, and who will suddenly become a genius struck by lightning.

I looked for common traits and found them in ambition and curiosity. When I told my American editor, with whom I had already worked with on a book about Edward Teller, about this book idea, she said it was very good, interesting, but that it would get boring after one or two or five cases. Let's also look for the unique characteristics of explorers. That's how I filtered out the 15 discoveries and explorers that eventually made it into this volume. They all shared, in common, ambition and curiosity, but each had a unique characteristic that was not repeated in the other 14. And indeed, the book became much more exciting as a result. There was one more thing that was a challenge: to give valid conclusions even for a reader who may not be interested in science. By the end of the book, it occurred to me that we can be most successful in any field if we find the activity in which we are good and can excel. This applies to any activity. It's worth thinking about and considering it because so many of us are not doing it day after day. What great things could be produced if most people could be dealing with what they were really the best at.

It requires self-knowledge, and that is the rarest commodity in the world.

I had a teacher who encouraged us to recognize the limits of our abilities. I feel I succeeded. My first favorite subject was mathematics, in some areas of which I was very good, but I realized that in other areas I was completely average. If you are ambitious, you should not go into a field with average abilities. I won a math competition, and got a chemistry book as a reward, and at the age of 11 I decided that that was my field, but I could feel the limits of my abilities there, too. It took me a long time to find what I wanted to do within chemistry. It's a huge area, all science is a huge area. Today science is so big that it has something new to offer to everyone, even to the narrow specialist within their narrow specialization. If you open a journal, even if you work in that field, you don't understand a good portion of the articles. Eugene P. Wigner, one of the world's greatest physicists, told me that he doesn't understand most of the articles in the *Journal of Chemical Physics,* not even the titles! I was shocked at first. It should be added, of course, that Wigner was very modest, and even exaggerated modesty.

Among Hungarians, for example, where do you rank the mathematician Pál Erdős, among the geniuses or "just" among the great scientists?

I just read an article about how birds see everything from above, they see the big picture, but they don't care about the details; frogs don't necessarily see the big picture, but they work out the details.

A witty observation.

If you look at it from this perspective — and this is not a value judgment, only a characterization — Paul Erdős should be classed with the frogs, and so should the great John von Neumann, along with the other great Hungarian physicists. The famous generation of Hungarian scientists, including the 5 Martians[1], were more frogs than birds. Both Erdős and von Neumann solved problems that came into their hands, but they did not really initiate them, their research did not follow from a grand overview but solved problems as they arose. In the case of Leo Szilard, we must note a very important difference. It is most interesting to compare Szilard with one of the greatest physicists of the 20th century, Enrico Fermi, with whom his career converged at several points. Fermi lived for science, whereas Szilard did science to save humanity. That is a big difference. Szilard was also a writer, with *The Voice of the Dolphins* being his best-known work. Michael Polanyi once told Szilard that he would be remembered for his writings rather than his scientific discoveries. Science is a huge edifice into which everyone brings their bricks, but sooner or later everyone becomes anonymous, with very few exceptions. However, writing a novel, composing a symphony, or painting a picture, is an individual achievement, which can only be done by the author and no one else. Michael Polanyi was a great scholar, and there are books of his that no one else could have written but him.

Is it like how you went from research to the history of science?

I remained a researcher all along, only the field and the focus changed. Before I started researching, I had a very broad range of interests.

---

[1]The five Martians of Science: Theodore von Kármán, Leo Szilard, Eugene P. Wigner, John von Neumann, and Edward Teller.

Then I realized that I couldn't create my research area with such broad interests, and my interests narrowed. For thirty years I was only interested in my own special field, the structure of molecules. Anyway, even today I am not a historian of science, I am just interested in the history of science, and I am mainly interested in the nature of scientific discovery, especially how one becomes a discoverer. You can learn a lot from the history of science. A lot of things are discovered that have been discovered before, only to be lost in the depths of the vast scientific literature. There is a great deal to learn from the history of science and not just from the successes — because they tend to appear in the scientific literature, but just as much from the failures.

The history of science helps to inform.

I find dissemination of knowledge useful for two reasons. One is that it is a great school for those who are disseminators. You can only spread knowledge if you understand what you are talking about. The other is that you are spreading knowledge to those who will probably have a greater influence on science than you, including those who fund scientific research.

In 1986, I was a visiting professor at the University of Texas, and I was asked to give a science course to students who were not going to become scientists and technologists. In America, they are required to take a course at the beginning of their university studies. As a lecturer, I was able to create the topic for this one-semester course myself and chose symmetry as the motif for the course. This subject is not far from the interests of non-scientists and brings them closer to the natural sciences. I thought I could easily manage this task without any special preparation. Fortunately, I soon realized that I was probably teaching the most important course of my life. For among my students were the lawyers, economists, politicians, MPs, military leaders, bankers, artists, critics of the future, and the people who would directly determine, for example, the laws by which we live and work.

Of course, we need to extend this realization to the whole of society, because only a well-informed society can make decisions in its own interests. We can say this more loftily, that democracy is inconceivable without the population being well informed, which includes knowing what scientists are doing, where science stands at any given time. It is often stressed that most scientific discoveries can be used for good and bad purposes, and scientists are no longer responsible for this. Scientists make discoveries, but it is society that decides how these discoveries are used. The responsibility of scientists stops there. But it does not stop there because society expects scientists (in a democracy) to inform society in a credible and understandable way. In line with this recognition, I believe that the dissemination of scientific knowledge is just as important as scientific research itself.

13

KRISZTINA PÉCSI, CATHOLIC RADIO BUDAPEST, 2014, MAY 13, "CONVERSATION WITH ISTVAN HARGITTAI ABOUT HIS NEW BOOK, *BURIED GLORY: PORTRAITS OF SOVIET SCIENTISTS.*"

Used with permission from Krisztina Pécsi

Who did you write this book for, what was your motivation?

There were several motivations. One was that the last time I was in Moscow, I visited the Novodevichy Cemetery again, where statesmen, artists, and generals are buried, and many great scientists. What struck me was that these scientists, who are otherwise little known to the wider world, were also great men, and at one time in the Soviet Union, they were very highly respected. Then there were periods when they were prosecuted. So, this visit was one of the motivations. Another was that for many years I've been trying to pass on scientific knowledge in a digestible form, and a book like this is an opportunity to do that. These scientists have made discoveries that affect our lives today and that was another motivation. Still, another motivation was that I was able to identify and select some of the most outstanding scientists, some of whom I knew personally.

What kind of people were they?

Of course, you can't generalize from 12 cases, but my overall impression is that they knew a lot, but they didn't flaunt their knowledge. They knew a lot, but they felt the limits of their knowledge. This made it even more exciting to communicate with them. Looking at them from a distance, you wouldn't notice anything special about them until you started talking to them. Then it all became an exciting adventure. The scientists chosen were almost all physicists. I'm a chemist, but I'm just as close to physics. In the twentieth century, a lot of physicists turned to chemistry and then to biology. One of my "heroes," Professor Semenov, whom I met here in Budapest in 1965, chose to become a physicist because he had read as a child that the great mysteries of chemistry would be unraveled by physicists.

What was the relationship between the twentieth-century systems of power and scientists?

This is a fascinating question, and it was one of my motivations to consider it. I was interested in how scientists who wanted to create and were able to create would fare under a totalitarian regime like the Soviet regime? The twentieth century was an exciting century for science, but it was also a terrible century for Nazism and for the Soviet system — I deliberately don't use the term communist system because it was such a degenerate version of the communist ideology that I prefer to call it the Soviet system or the Stalinist system. These totalitarian regimes were all anti-scientific, even if they did give privileges to certain scientists during certain periods. There was also a selfish reason for this from the point of view of the system. The question can also be asked: Is it possible to create in a totalitarian system? Creations have been made in all totalitarian regimes because certain talents can no longer be held back above a certain level.

Ernst Neizvestny (left) and Istvan Hargittai, 2014, in Neizvestny's Manhattan home studio, by Magdolna Hargittai.

Recently, in New York, I visited the Russian sculptor Ernst Neizvestny,[1] something I had wanted to do for a long time. He is famous not only for his modern sculpture but also for his confrontation with the then Soviet leader Nikita Khrushchev at a major exhibition in Moscow in 1962. Khrushchev was terribly critical of modern art. This included his intention to please his critics, including communist party leaders, who blamed him for introducing a measure of liberalization, and he wanted to demonstrate that he was not going beyond Soviet dogma in all things. He went to this exhibition and there he was merciless in his criticism of modern art, including Neizvestny's work. Neizvestny was brave and confronted Khrushchev. Of course, this was not a debate between two equals. For years afterward, Neizvestny received no commissions — state commissions, but there were practically no other commissions at the time. He was exiled from the Soviet Union in 1976 and eventually became a very

---

[1] Ernst Neizvestny (1925–2016).

successful artist in New York. He was a good subject to ask: how can you create in a totalitarian regime? He said that one must create, as simple as that. There is a creative talent and a creative urge that no system can hold back.

Soviet scientists worked under difficult, often impossible conditions. Why were they loyal to the system?

There are several factors to consider. Let's take a striking example. In World War II, Nazi Germany, which attacked the Soviet Union, attacked not simply the Soviet system but the very existence of the Russian people. Not only out of political considerations but also out of patriotic considerations, indeed out of pure patriotic considerations, Soviet scientists in the service of the Soviet state did everything in their power to ensure that the Soviet Union defended itself and defeated Nazi Germany. So far, the matter is clear. The question can be continued: why did they work so hard to create the atomic bomb and the hydrogen bomb after World War II? They were isolated and could only see, perhaps apart from one exceptional scientist, that the Soviet Union was threatened by an enemy even more powerful than Nazi Germany, and they did everything they could to ensure that the Soviet Union could counter the American nuclear threat. They created the Soviet atomic bomb and developed the Soviet hydrogen bomb. There were also some who only took part in the work out of necessity because they feared for their lives and could have met the saddest fate if they refused to work. It was Lev Landau, who later won the Nobel Prize, who stopped working on nuclear weapons immediately after Stalin's death. The others, however, continued and only later realized how lucky the world was that Hitler and Stalin had not been the first to have nuclear weapons with which to blackmail the democratic world.

Stalin's terror extended to scientists, but there were brave ones, like Peter Kapitza.

Kapitza was stripped of his position in the post-war period. Indeed, before the war, he was very brave in standing up for persecuted scientists, most of whom were arrested and imprisoned on trumped-up charges, including Lev Landau. There were two distinct phases in the Stalinist terror. The first was in the second half of the 1930s, when Stalinist paranoia and the quest for autocracy created a terrifying terror and indiscriminately considered everyone as an enemy, military leaders, political rivals, scientists, and hundreds of thousands of ordinary people. A lot of people were executed, and they didn't have to be a real enemy to meet such a fate. After World War II, in the last years of Stalin's reign, there was another wave of terror, but with a very strong anti-science and anti-Semitic component. The Soviet dictator had terrible plans to destroy the Jewish anti-fascist movement, starting with its leaders ...

Was Stalin an anti-Semite?

We don't know if he was originally, but after the war, in his last years, he pursued the bloodiest anti-Semitic policy and wanted to deport the Jewish population of the European part of the Soviet Union to remote Siberian territories. Only his death prevented him from doing so. Several of the scientists in my book were Jews, and those who worked on the development of nuclear weapons were able to survive thanks to their work. Stalin said let them work, we could execute them later. Stalin's paranoia applied to everything that came from the West. It made him anti-scientific. They launched huge attacks on modern biology, which they crippled and destroyed for decades. Cybernetics, now more commonly known as information technology, was seen as bourgeois science. In chemistry, too, modern approaches to research were attacked. They were already preparing to take ruthless action against physics because quantum mechanics and relativity were also considered part of the Western bourgeois ideology. At the time, the physicists said that either they could apply quantum mechanics and relativity theory and then they would create Soviet nuclear weapons, or else there would be no Soviet nuclear bomb. Physics was then given a grace period until the nuclear weapons were ready. In the meantime, Stalin died, and the Soviet physicists were saved. When the first Soviet experimental nuclear explosion happened in 1949, two scenarios were prepared for it. If the experiment succeeded, hundreds of scientists would be honored, ranging from the highest honors to progressively lower levels of honors depending on the importance of their participation. However, in case of failure, those who would have received the highest honors would have been executed. Those who would have received lower honors would have received severe and progressively less severe prison sentences.

Many of them later became Nobel laureates.

Which, of course, they did not get from the Soviet system. Occasionally though the Soviet system had a hand in that too, and that was after Stalin. For many years it was not possible to send nominations from the Soviet Union to the Nobel Prize committees. The Swedes, on the other hand, tried to involve Soviet scientists in what they called the Nobel movement, and they were willing to compromise. The nominations are strictly secret, and no collectives, organizations or institutions are allowed to participate in the nominations. However, in the case of the Soviets, the Swedes knew that the Soviet nominations involved the Academy of Sciences, the Foreign Ministry, even the secret police, and of course, above all, the Communist Party. The fact that they won Nobel Prizes in physics and chemistry was because they did have great scientists who made important discoveries despite difficult circumstances and international isolation. If science were isolated in Hungary, the consequences would be tragic. In the case of Soviet science, the consequences were milder, because the internal scientific life of this huge country was strong enough to mitigate the consequences of international isolation to some extent. Even so, isolation did enormous damage and, likely, prevented important discoveries. There were prominent Soviet scientists who marked on the walls of their office a map of the world where they had been invited but not allowed to travel. When the Soviet Union fell apart, many of them left and never returned. I would add that, to this day, science in Russia has not been fully integrated into international science.

Can we no longer expect a similar performance from scientists in Russia?

Funding for scientific research has been significantly reduced. And there is another important reason for this. In the Soviet Union, there were very few tracks in which there was at least some freedom of movement, some room for autonomy. These included sports, music, and science. Today, talented young people can go into business, advertising, financial services, foreign trade, and lots of places. And many go abroad. Far fewer of the talented young people who choose a career in science stay. We cannot expect a performance in science in Russia today comparable to that of the Soviet times.

Are any of the scientists in this book still alive?

Alexei Abrikosov is still alive, he has emigrated to America, he is in his mid-eighties, and he is seriously ill.[2] He was very successful at the time and was in his late 60s when he moved to America and became successful there, too. He was awarded the Nobel Prize, as an American, but for his achievements in the Soviet Union. We visited him a few years ago, he lives in a nice family house with his wife. I had a sad impression of him. Everything in the house was Russian, and I don't know if he doesn't think he might be more comfortable in Moscow, but he certainly wouldn't admit it.

---

[2]Alexei A. Abrikosov (1928–2017), Russian American physicist, Nobel laureate.

14

JEREMY N. A. MATTHEWS, "QUESTIONS AND ANSWERS WITH ISTVAN HARGITTAI." *PHYSICS TODAY* 2014, AUGUST 28. Reproduced with permission of the American Institute of Physics.

A research scientist turned prolific biographer explains his fascination with the great minds in science.

Even with his accomplishments as a research chemist, Istvan Hargittai may be more broadly known for the books he's written about the lives of famous scientists. The list of his subjects includes James Watson, Lev Landau, Peter Kapitza, and, as he calls them, "The Five Martians of Science" — Hungarian physicists Leo Szilard, Edward Teller, Theodore von Kármán, John von Neumann, and Eugene Wigner.

Himself a Hungarian, Hargittai earned his doctoral degree from the Hungarian Academy of Sciences. In 2012 he became professor emeritus at the Budapest University of Technology and Economics, where he has taught for more than two decades. To keep current, he says he still lectures and edits the international journal Structural Chemistry.

Hargittai's most recent book is Great Minds *(Oxford University Press, 2014). Written with his son Balazs and his wife Magdolna, who are also chemists, it features the stories of 111 living scientists. His other biographies include three that were reviewed in Physics Today:* The Martians of Science: Five Physicists Who Changed the Twentieth Century *(Oxford University Press, 2006),* Judging Edward Teller: A Closer Look at One of the Most Influential Scientists of the Twentieth Century *(Prometheus Books, 2010), and* Buried Glory: Portraits of Soviet Scientists *(Oxford University Press, 2013).*

Physics Today recently caught up with Hargittai to find out what drives his passion for writing about scientists and their work.

At what point in your career — and why — did you make the switch from research chemist to biographer and historian of science?

There was no clear-cut switch. Even at the start of my scientific career in the mid-1960s, I was interested in writing, but I soon realized that I should focus on my niche of research — molecular structure and modeling, using primarily electron diffraction. In 1969, I was a visiting research associate at the physics department of the University of Texas at Austin, and a weeklong visit by Eugene Wigner had an important influence on me. I had been in correspondence with Wigner from my student years, and now he spent an hour with me every morning and expanded my knowledge of and thinking about symmetry.

That interest in symmetry made me write for *The Mathematical Intelligencer*, and eventually, it triggered the founding of a magazine about the culture of chemistry, *The Chemical Intelligencer*. For the magazine, I interviewed famous scientists. This activity continued beyond the magazine and included chemists, physicists, biomedical scientists, materials scientists, and mathematicians.

My wife, son, and I together published a six-volume book series, *Candid Science* (Imperial College Press, 2000–2006). These volumes contain in-depth conversations about the scientists' lives and their science. *Great Minds* is a collection of excerpts from 111 of our interviews, and it is for a broad readership without any prerequisite of science background. The transition you asked about, from research chemist, was gradual and completed only when I became emeritus in 2012.

You discuss what drives great scientists in your book Drive and Curiosity: What Fuels the Passion for Science *(Prometheus Books, 2011). What has driven you over the past two decades to meet, interview, or write about them?*

Ever since I received my first science book — as the prize for winning a math competition when I was 11 years old — I was interested not only in science but also in the discoveries, the history of the discoveries, and the careers of scientists. I believe my interviews have been successful because my questions originated from my genuine interest, and I was equally interested in the human fate. These interviews made it possible for me to learn about areas of science in which I felt ignorant and "my teachers" were the best minds.

Many areas in the biomedical sciences were examples. My interviewees understood that the person who is asking them all those questions is a fellow scientist and not a journalist, and they often opened to me to the extent that they were later surprised by the depths of their own responses. I consider the interviews, which were a family project, as my second university education. It was then quite natural that I did not want to keep what I learned to myself but wanted to share it.

What is it about the scientists in Great Minds *— and about the profession of science in general — that you'd like to impart to the next generation of scientists and to the general public?*

For the next generation of scientists, *Great Minds* should induce them to learn more about these scientists and perhaps look up the *Candid Science* volumes for the in-depth interviews. For the general public, there were two motivations to create *Great Minds*. One was to bring these great scientists into human proximity, to share their interests, struggles, and views on a diverse set of issues, and to show that they are quite ordinary people except for their great achievements in a narrow area of human activities.

The other motivation was to convey the impression that from the careers of successful scientists, we can learn a great deal beyond science. The discoverers demonstrated that to accomplish the most, the best one can do is what one is best at doing. However trivial that sounds, are there many people who are engaged in doing what they like and can do best?

What so far has been your favorite — and least favorite, or most difficult — book to write?

People who look at my output think that I write easily. This is not quite so. The first draft is always the most difficult, and I keep rewriting, often many times. Each time I embark on a new book I find it very challenging. Looking back, my semiautobiographical book *Our Lives: Encounters of a Scientist* may have posed the greatest challenge. If a fire were to destroy all my books save one, I would probably want it to be *Our Lives*. It is one of my least known books.

Are you currently working on another book?

I can tell you about a book at the start of production. It is coauthored with Magdolna: *Budapest Scientific: A Guidebook*, to be brought out by Oxford University Press. It is about visible memorabilia of science and scientists in Budapest. It will be quite a challenge to produce it with nearly 800 images. There is history, there are stories, even some politics in the book, but we believe it is much broader than just one city. The science in Budapest and especially the scientists from Budapest have affected world science and culture more than their home country.

What books do you like to read and which ones are you currently reading?

I read a lot, mostly nonfiction and often about topics related to my writing, but not necessarily for my writing. Currently, I am reading Ernst Neizvestny's book *Bella Dizhur* (2013). Neizvestny is a great Russian sculptor, and this book in Russian is about his mother, Bella Dizhur. Neizvestny was a dissident avant-garde artist in Moscow who produced Lev Landau's tombstone. He now lives in New York, and last fall my wife and I visited him in Manhattan. He gave me this precious book about his mother.

15

CSABA MOLNÁR, "CRACKS IN THE PAVEMENT" (EXCERPTS). *MAGYAR NEMZET*, AUGUST 27, 2016 (DAILY NEWSPAPER).

Used with permission from *Magyar Nemzet*

*W*isdom of the Martians of Science is the title of a new book by chemist Istvan Hargittai, the tireless chronicler of 20th-century Hungarian scientists (co-authored by Balazs Hargittai). The book tells the world-shaping nature of science through the life stories and quotations of five Hungarian Jewish researchers who emigrated to America.

Do you consider yourself a scientist or a historian of science? Does it bother you that the public knows you better as a biographer of other scientists than as an independent researcher?

I am a researcher, and these activities cannot be strictly separated. The history of science is a discipline and needs to be studied. I am a complete amateur at it. I am interested in the history of scientific discovery, but even more in the nature of discovery. More people are aware of my books that reach many thousands of readers than of those that reach only a few hundred specialists.

How did you become interested in the life stories of other scientists?

I started my university studies at Eötvös University in Budapest in 1959 and completed them at Lomonosov University in Moscow in 1965. Right at the start, Géza Simonffy, the legendary popular science editor of Radio Budapest, asked me to interview the Soviet Nobel laureate Nikolai Semenov, who was visiting Budapest. The Radio gave me a technician and a recording device the size of a small suitcase. I visited Semenov in his room at the Hotel Royal. I thought I knew everything I needed to know to do an interview, but I knew nothing. The only reason this didn't turn out to be a disaster was that Semenov was an experienced interviewee. The interview was broadcast several times on the radio and was published in a volume of the best of radio programs. Thirty-five years later I bought a copy of the recording from the radio archives and listened to it, I liked it, even though, by then, I was an experienced interviewer.

Did you learn interviewing quickly?

After Semenov, I did a few more interviews for radio, but none of them were particularly interesting. I stopped doing this anyway because I was concentrating on my research. I was living my work with such a passion that I could even see the curves of my measurements in the cracks of the pavement. Then, about 25 years later, my interests began to widen again. I started publishing articles in a popular mathematics magazine that were not specifically related to my research. I liked the magazine so much that I suggested to the New York branch of Springer Verlag to publish a similar magazine in chemistry. As the editor of this new magazine, I started interviewing again, this time famous scientists, chemists, physicists, biomedical scientists, and others. In time, my wife and then our son Balazs joined in. For me, these interviews were like a second university. I have never been as unprepared for an interview as I was when I did it for the first time.

Most of the Hungarian scientists who emigrated before the war were of Jewish origin. Do you think religion and origin have a role in what they achieved, or could others have achieved something of the same caliber if they had been the ones being persecuted?

I am not an expert on this question, but there are considerable studies on the subject. There are many factors that have led to so many Jews being among eminent scientists. The traditional respect of Jews for learning and knowledge but also persecution and discrimination, may play a role. In the mid-1950s, like me, several of my schoolmates were not admitted to high school. It was because of my late grandfather who was a business owner, and the others because of their kulak [well-to-do peasant] origins. In time, however, we all graduated and did very well. The barriers that were supposed to prevent us from studying stimulated us to perform better. And as for the impact of emigration, when you are in a new country, in a new environment, it is a powerful incentive to try harder. It is also true, of course, that we tend to hear more about successes and less about failures.

Have you ever thought about, following other scientists, emigrating?

Yes. When we were in Texas in 1969, my wife and I asked ourselves that question. But we decided to return home. If I tried to explain why, it would sound too pathetic.

Many people are proud of the many national achievements of the achievements of Hungarian scientists who have become world famous. Others, however, think we should be ashamed that these people had to flee to succeed (or merely to survive). Which camp do you belong to?

You can be proud, and you can be ashamed. What you should not do is obfuscate the reasons why these excellent people left. One should also face the tragedy that the 25 years of the Horthy regime (between the two world wars) represented for science. During a quarter of a century, hardly any Jewish scientists were allowed to be appointed as professors and elected as members of the Academy of Sciences. And then there was the *numerus clausus*,[1] which excluded many Jewish young people from higher education, and which Bálint Hóman, the leading cultural and educational politician, sought to extend to sec-

---

[1] Hungarian anti-Jewish legislation in 1920, the first such anti-Semitic law in post-World-War Europe, severely limiting the number of Jewish students in higher education.

ondary schools. Until we face the fact that the Horthy regime played a major role in preparing for the Holocaust, it will not be possible to answer your question fairly.

In your student days, in the fifties and sixties, were physics and chemistry more highly regarded than today?

In a country as balanced as Britain, there is very little fluctuation in the appreciation of knowledge. In the United States, the Soviet Sputnik gave a huge boost to the appreciation of science and science education, in particular. In the Soviet Union, defense objectives dictated a special appreciation of the natural sciences, especially physics. As far as present-day Hungary is concerned, the drastic reduction of science education, and education in general, that goes against the creation of a knowledge-based society, is shocking. I feel as if we are part of a bad dream and there is no one to wake us up.

Isn't part of the reason for the decline in the quality of education that today's students have lower skills?

The level of knowledge of first-year university students is probably lower than when I became a university student. Then, for many, including me, it was so difficult to get into university that it encouraged excellence. But as far as the talent of students is concerned, it doesn't change, the question is how well you manage to develop that talent. My many years of experience as an educator in the United States tell me that we should stop trying to achieve a uniform level of education and allow for differences in quality. One consequence of this would be that employers would not only be interested in what the employee graduated from but at least as much from which university. As with many other things, competition could have a positive effect here too.

Are there future Martians among today's students?

This is a label to be treated with care. By my definition, the "Martians" were the great scientists who risked their careers in science to defend the free world and democracy during the Second World War and the Cold War. I agree with Edward Teller, who says that Theodore von

Kármán, Leo Szilard, Eugene P. Wigner, John von Neumann and he, Teller, were "the Martians." Others say that the Hungarian emigrants who became successful in the West were all Martians. Sticking to the narrower definition, I do not think it likely that there were any other Martians, because it took a very large number of circumstances to make the careers of these five individuals come together as they did.

You wrote a separate book about Edward Teller. The perception of his role in the Cold War is not uniformly positive in America.

The Martians were not easy people. They were not always understood by their great colleagues, and on several occasions they themselves described each other as not understanding each other. My book on the five Martians led me to write a biography of Edward Teller. There had been previous biographies of Teller, but all the previous books had either been unconditional adoration or blind hatred. Teller really polarized people. It was universally acknowledged that he was an extraordinary physicist, but his political activities were controversial, and, on occasion, his human behavior was despicable. My aim was to write a balanced biography.

What was your personal impression of Teller?

Our conversation got off to a difficult start, he clearly didn't have a desire to repeat things we knew. His wife warned him not to be so unfriendly. I remember what changed his mood. Many people attributed Teller's anti-communism to the fact that he had lived through the ordeal of the Hungarian Soviet Republic in 1919. But Teller was only 11 years old at the time, and he spent the better part of the 133 days of the communist dictatorship not in Budapest, but in Lugos, a small town in the southeast and far from Budapest, with his grandparents. Fortunately, I knew that his political thinking had been greatly influenced by Arthur Koestler's *Darkness at Noon*. When I mentioned this, Teller realized that I didn't want to reduce our interview to clichés, and from then on, I could ask him about anything. The way Teller explained the Heisenberg Uncertainty Principle was a great experience; even then I felt it was unique. It was as if we were experiencing it on a different level from our everyday existence,

With Edward Teller in the Tellers' home in Stanford, California, 1996, by Magdolna Hargittai.

and it was only when the interview was over that we were back on solid ground. Apparently, Teller was also sorry that the interview had ended, even though it was one of our longest recordings, which came out of that encounter.

Have many Nobel Prize-winning interviewees turned out to be terribly boring people in real life?

One of the lessons from our interviews is that a Nobel Prize does not make someone great who was not great. The Nobel Prize is not awarded to the greatest scientists, it is awarded for the greatest discoveries. A Nobel laureate can be just as petty and narrow-minded as he can be broad-minded and interesting. A Nobel laureate can also be a fool but finds something that is a milestone in science even if the discoverer does not even recognize its true importance. What is certain that every Nobel laureate has achieved something for which he or she deserves respect and thanks. The mistake is not made by the Nobel laureates, but by us, when we assume that once they have been awarded the Nobel Prize, they are not only outstanding in that one thing, but in everything else.

16

ZSUZSA KUN, KLUB RADIO BUDAPEST, JULY 11, 2020, FOLLOWING THE PUBLICATION OF *MOSAIC OF A SCIENTIFIC LIFE.* EXCERPTS FROM THE CONVERSATION

Used with permission from Zsuzsa Kun

Great scientific or artistic achievement is not always matched by attractive human qualities. An example is Edward Teller.

Meeting Teller was a great experience. The conversation started slowly, then at a certain point, the mood changed. I have often found in similar conversations that there is a point that must be reached somehow, that turning point has to be reached for the conversation to change into an exciting experience. There is no general recipe for that. In this case, the turning point was when Teller understood that I knew he was not an anti-communist from childhood, as many people had assumed, but that his views had evolved by his later experiences. At the age of 11, there was a Hungarian Soviet Republic here, but that did not shape Teller's worldview, nor was he in Budapest for most of the 133-day dictatorship. Koestler's book *Darkness at Noon*, published in 1941, played a role in forming Teller's worldview. More generally, Teller's anti-communism was not so much ideological as it was political. He wanted to defend the United States.

At the Oppenheimer hearing, Teller spoke against Oppenheimer and was given a black mark in America.

Among the physicists, but even in America, not everyone felt that way. In political and military circles, his credibility was boosted. We now know that Teller's testimony was not decisive for the outcome of the Oppenheimer hearing. However, it was humanly undignified, and he put it so cleverly, one might say so cunningly, that it led to a particularly long-lasting hostility toward him among his peers. Some of his colleagues would not even accept his hand for a handshake.

And the Hungarian perception?

All we saw here was that he was a great physicist, reviled by the socialist system, so he must be good. Following the political changes, Teller visited Hungary and gave inspiring talks. Here his acclaim was almost total. Among the exceptions were environmentalists who recognized the damage he could have done. But everyone knew he was an important man. Looking a little further back, I think there were four moments in the twentieth century when Hungary stepped onto the world stage. These four moments are, to put it symbolically, Béla Bartók's music, Ferenc Puskás's football, Imre Nagy's politics, and Edward Teller's physics.

I thought you were going to mention four Nobel laureates.

Teller and the hydrogen bomb had world-historic significance.

What do the lives of Nobel Prize winners tell you? It was not clearly a success story; you could say they had their crosses to bear.

Or their Star of David. Their fate says to me, first and foremost, that those who became successful had to leave Hungary. That is a very sad lesson. The problem is not that scientists leave, but that this movement is one-way.

One of your articles criticizes the way the history of the Hungarian Academy of Sciences is presented on its website. You write, among other things, that "Instead of confronting the past, the document on the website of the Hungarian Academy of Sciences is a falsification of history, akin to the House of Terror and the Freedom Square memorial to the German occupation.[1] It is unsuitable for giving us a realistic picture of the situation of Hungarian academia in the Horthy era, and for revealing that for twenty-five-years there were anti-Semitic laws affecting Hungarian academia, there was a Hungarian Holocaust in which Hungarian academia suffered losses, there was a world war, and we cannot even get a realistic picture of how discriminatory academic elections were in the Horthy era."

You also quote Albert Szent-Györgyi's sharply critical words of November 30, 1945, concerning the Academy's role in the Horthy era: "... I have noted with sincere and deep regret that the Academy has closed itself off from any reform that would have led to its renewal. Without such a renewal, the Academy cannot be a worthy representative of Széchenyi's ideal. The Academy is largely responsible for our national disaster. According to the vision of its founder, the Academy should be a citadel of intellectual independence and progress.[2] Instead, the Academy has become a nest of servility and pseudo-science, and this Academy, which found in [Hapsburg] Archduke Joseph its most worthy leader and in

---

[1] House of Terror is a biased museum in Budapest, which presents the crimes of communism and the Hungarian Nazis (the Arrow Cross movement) but is silent about the crimes of the anti-Semitic Horthy regime. The memorial on Freedom Square remembers the German occupation of Hungary and implies that all anti-Jewish crimes, all crimes against humanity during World War II, should be ascribed exclusively to Germany.

[2] István Széchenyi (1791–1860) was one of a small group of Hungarian aristocrats who founded the Hungarian Academy of Sciences. It was only one of many of his creations and suggestions toward modernizing Hungary. He was called "the greatest Hungarian."

Orsós[3] *its scientific ideal, still sits together almost unchanged and obstructs the path of the country's rebuilding."*

After my repeated protests, I only managed to get two or three sentences added to the description of the history of the Academy on its website. The response to my first criticism was that if the facts I had missed were mentioned, the whole document would have to be rewritten so as not to alter its proportions.

Why did you not leave Hungary? In your book, you mention that from the very beginning, your life was a struggle, although you then became successful.

Why I didn't leave, or why we didn't leave, I should rather say, because I didn't decide anything on my own, my wife and I always discussed everything and decided everything together. The first time we were together in the United States, in 1969, I for one year and my wife for three months, we asked ourselves this question: should we stay or should we go home? Two things were important. We were in Texas, and it was not attractive to us. Maybe somewhere in the Northeast would have motivated us a little differently. The other thing, and I say this without any embarrassment or shame, we were enthusiastic, we wanted to create science at home, and we did create it, under very difficult circumstances. I cannot say today that we have regretted our decision.

Your children are not in Hungary.

We have a son and a daughter, and I can honestly say that we are glad they are not here.

---

[3]Ferenc Orsós (1879–1962) pathologist, member of the Hungarian Academy of Sciences (1928–1945). He was a vocal anti-Semite, and he was directly involved in the deportation and murder of 2500 Jewish physicians. He was excluded from membership of the Hungarian Academy of Sciences in 1945. From December 1944, he lived in Germany. The allied powers denied his extradition to Hungary despite repeated requests.

They were born into a better world.

We didn't really influence them, but when they asked us, we said that they could have an easier, nicer, and more perspective life where they were going. At first, they both lived in America, got their PhDs, and became U.S. citizens. Our son is a chemistry professor and has a family. Our daughter is a professor of sociology and communication sciences at the University of Zurich. We have an extremely close relationship with both and are happy to have them where they are.

Who are some of the people you've become friends with?

There have been times when I was slow to realize that I had formed a real friendship. I was talking to a professor in California and then we met a few years later, and he almost verbatim told me all the things we had talked about and how it had affected him. That's when I realized that friendship is not just about how much time we spend together but also about how we affect each other.

How much does a Nobel Prize change a scientist? Is there even a need for such recognition today?

It is not necessarily beneficial for the individual, even for those who receive it. For those who come close and do not receive it, it is often a tragedy. For science, for humanity, it is a very important factor, because at least once a year everyone's attention turns to science, and the Nobel Prize becomes a subject of public curiosity. Perhaps we should be a little more realistic about its importance and not think that a Nobel Prize winner can do everything at once and that whatever he says should be treated as a revelation. He is not good at everything, but he is certainly very good at one thing. As for the impact on the individual, it may be very different. George A. Olah, for example, has hardly changed. Fortunately, I encountered him before his Nobel Prize. I used to say, with a smile of course, that if you want to be on good terms with a Nobel Prize winner, you should contact him a year before the Nobel Prize, because then this interest obviously does not come from his world fame. Olah made a conscious decision to continue to devote his life to scientific research after a short break for celebration, and he has succeeded in doing so. I have had friends who have won Nobel Prizes who could not get over the fact that they had won the Nobel Prize and over the years have repeatedly wondered why they had won it and not someone else who could have just the same.

With James D. Watson (right) in his office at Cold Spring Harbor Laboratory, 2000, by Magdolna Hargittai.

James Watson, the discoverer of the DNA double helix, sold his Nobel Prize for $4 million and got it back, but he couldn't have known that in advance.

I must refer to him as Jim Watson, I have become close to him over the years. He was a controversial figure whose friendship was not originally due to me, but to his great respect for Hungarian intellectuals, the so-called Martians, especially Leo Szilard, and also George Klein, with whom Watson was on good terms. I was also good friends with Klein. Watson somehow transferred his respect for Szilard and Klein to my wife and me. When I first wanted to interview him, he wrote back that he was unavailable, but then he contacted me on his own. By the end of our conversation, we had grown so close that within six months he and his wife visited us in Budapest, and he invited us to stay for three months at his Cold Spring Harbor Laboratory, which turned out to be a wonderful experience. I wrote my first autobiographical book largely as Watson's guest.

So, why did he sell his Nobel Prize?

Of course, he didn't sell his Nobel Prize, but only the medal, of which he had two copies left. He could not have known in advance that it would sell at auction for $4 million, which could have been half a million, which is still a lot of money. The Watsons have two sons, one of whom is manic-depressive, and Watson wants to care for him as best he can. This was Watson's explanation for putting the medal up for auction. It had been sitting in a display case where few people could see it, or worse, in a bank vault where no one could. It is the copies that are usually on display. You can't tell the difference between the original and the copy. Of course, he could not have expected to be bought by a Russian billionaire who believes that Watson should not be deprived of his Nobel medal. So, he decided to give it back to Watson. The return ceremony took place in Moscow at the Russian Academy of Sciences, where the President of the Academy presented Watson with the medal.

Watson is a tragic researcher. His discovery with Francis Crick was based in part on information from Rosalind Franklin and her colleague that was not entirely clear how they had obtained it, so this source was not even acknowledged in the announcement of the discovery. Watson was always an eccentric, but perhaps he preferred to be seen as an eccentric. Towards the end of his career, he made some scandalous statements, some seen as anti-feminist and some others as racist. Eventually, he excluded himself from the scientific community, and, above all, he was distanced and disowned by his own research center. This research venue, the Cold Spring Harbor Laboratory, already existed when he became its director, but Watson made it a world-renowned hub of biomedical science. As a result of a car accident, he is now in a long coma. A sad end to a brilliant career.

We still have a few minutes; I also knew George Klein and you interviewed him. He is also in your book.

Two themes have come up between us, repeatedly, over the years. One is that we really begin to appreciate life when it is in danger. He dealt a lot with the question of death. I usually just listened to that. But what we talked a lot about was conformity. How we adapt to the powers that be, for example, or to the expectations of our social environment, for example, is a very interesting topic, for me as well. I have always tended to neglect these expectations. It has never bothered me, and it never bothers me, to be in the minority or even alone in my opinions at an academic class meeting, for example. The important thing is to be able to express my opinion. Conformism was a common interest and whether we agreed or disagreed, we understood each other very well on that.

George Klein in the Hargittais' home in Budapest, 2000, by Istvan Hargittai.

He was a very quiet, solid, reserved man.

He wasn't quiet at all, he was just soft-spoken; it was only his appearance that gave that impression, or tried to give that impression, and he was successful in doing so. Behind the quiet, solid appearance was a strong, almost aggressive, but in a good way, opinionated and assertive personality. He almost wore a mask, not against a virus, but against the preconceptions about him.

Was the Nobel Prize never in your secret ambition?

My secret desires never included the Nobel Prize, but George Klein's secret, even subconscious desires, rightly of course, may have included it, and that may have been the reason why he became so hostile to the institution of the Nobel Prize. For many years, he was a member of the assembly of leading scientists at the Karolinska Institute that decides on the Nobel Prizes in Physiology or Medicine, and then he was one of the most outspoken critics of the Nobel Prize. He was a very interesting, exceptional personality and it was a great pleasure to know him and to debate with him.

Does being a member of the committee that decides on the Nobel Prize exclude them from receiving it?

It does not exclude, and there have been several examples.

17

KRISZTINA PÉCSI, CATHOLIC RADIO, BUDAPEST, OCTOBER 30, 2020, CONVERSATION WITH ISTVAN HARGITTAI ABOUT THE NEW BOOK IN HUNGARIAN, EDITED BY HIS SON, BALAZS HARGITTAI, PRESENTING A SELECTION OF ISTVAN HARGITTAI'S NON-TECHNICAL WRITINGS

A similar collection of selected writings has appeared in English, Balazs Hargittai, Editor and Compiler, *The Culture and Art of Scientific Discoveries: A Selection of Istvan Hargittai's Writings* (Springer Nature, 2019). Excerpts from the interview used with permission from Krisztina Pécsi.

What is behind the title of this new volume?

You might think that I only deal with science, which is of course not true because I also deal with science-related education and popularization. I met with Eugene P. Wigner in 1969, and this meeting in Texas with this Hungarian-born Nobel laureate has had a strong influence on my interests outside science. This meeting had a personal atmosphere, a certain intimacy, and I am a bit like that with science.

During the first 35 years of your research career, you produced hardly any non-technical papers.

I tried to be at the international forefront of my own field, molecular structure research, and narrowed my interests. I forced myself to do this because previously my interests were too broad. This broad interest has come back after 35 years.

With Eugene P. Wigner (left) at the University of Texas at Austin, 1969, by unknown photographer.

You have had at least one conversation with hundreds of famous scientists. You also write about scientists you have never met in person.

For example, Leo Szilárd and Enrico Fermi. I have never met them, but I have read so much about them that I regret not being able to ask them a million questions about their fascinating lives. In one of my writings, I compare the two. They had many differences of nature, of opinion, of interest, yet when it came to working together to develop a weapon that the Germans could create — the atomic bomb — they put aside their differences and did their job brilliantly.

The story of the Martians spans an entire historical era and is of global historic significance.

We love to talk about how important Hungarians, including Hungarian scientists, are to the development of the world and science, but this is an exaggeration. But what is not exaggerated and cannot be overstated is the significance of the achievements of five Hungarian scientists, the so-called Martians, Theodore von Kármán, Leo Szilard, Eugene P. Wigner, John von Neumann, and Edward Teller. They were instrumental in the success of the United States in the war against Nazi Germany and militaristic Japan and then in the Cold War against the militarily very powerful Soviet Union. It is inconceivable that a situation could ever again arise in which five scientists could play such a huge role in such fateful events. The five of them came from roughly the same neighborhood in Budapest, from upper-middle-class Jewish families, studied at three different high schools, and were forced to emigrate, first to Germany and then to the United States. There they played a crucial role in the application of science for defense purposes.

One of the essays asks the question: could Leo Szilard have been wrong?

Szilard was famous for foreseeing events that later happened. It is therefore interesting to find a case where he was wrong in predicting the future. For example, he was wrong in his idea that nuclear technology would one day make energy so cheap that it could be treated like the air we breathe or the water we drink. We might add that today, access to clean air and clean water is unthinkable without sacrifice. He also imagined, as Edward Teller did, that nuclear explosions could be used for large-scale peaceful, nature-transforming developments. Such ideas have been expressed elsewhere, too. I learned as a schoolboy that in the Soviet Union, the course of the great Siberian rivers would be reversed to create thriving agriculture in areas where nothing had grown before. Szilard had similar ideas, but he was wrong. However, Szilard did not stubbornly cling to ideas which, for one reason or another, turned out to be impossible to put into practice. In fact, one of his most important lessons is to have the strength and courage to abandon a project if it turns out not to work, no matter how much time and energy you had invested in it.

With Vitaly Ginzburg (right), in Ginzburg's office at the Lebedev Institute of Physics, Moscow, 2004, by unknown photographer. Vitaly L. Ginzburg (1916–2009), Nobel laureate, provided one of the three basic ideas for the Soviet hydrogen bomb (the other two were from Andrei Sakharov).

One of your articles compares the first Soviet and American nuclear weapons laboratories.

There were at least as many similarities as differences. Let's look at an example of the differences. The Soviet scientists went to the secret nuclear laboratory under duress, and the Americans, including the Hungarians, took the initiative themselves. However, there was a similarity in that the American laboratory was staffed by many European refugee Jewish scientists. The reason why there were so many was that by the time they arrived in America and by the time they were able to get involved in secret projects, the Americans were already working on other subjects, such as the radar. That's why there were relatively large numbers of refugees involved in the nuclear program, and because many of them were pioneers in the field. What was the situation in the Soviet nuclear laboratory? In the Soviet Union, and few people know this, there was strong discrimination against Jewish citizens, and this was almost total in scientific research and its applications. The nuclear program, because of its importance and urgency, was an exception, and Jewish scientists were allowed to participate. This is why the proportion of Jewish scientists in the first Soviet nuclear laboratory was high. The nuclear laboratory provided a safety net for Soviet Jewish scientists, and in this, we can see a similarity between the Soviet and American laboratories.

You have recorded interviews with more than a hundred Nobel laureates. How did you get in touch with them?

In a variety of ways. I started a magazine on the culture of chemistry and these interviews played an important role in it. Readers were interested in these in-depth conversations, both professionally and personally. When the interviews were published by a London publisher, I remembered that long before my great interview program of the 1990s, I had already interviewed the Nobel Prize winner Nikolai Semenov in 1965 at the request of Radio Budapest, and that interview was included in the first volume of the six-volume *Candid Science* series in 2000.

You have been asked to nominate for the Nobel Prize every year for decades. How secret are the nominations?

They are absolutely secret, and I keep to that. It is not appropriate for a nominator to talk about it and if news of such nominations do leak out, it is not appropriate for a potential nominee to talk about it either. I do not talk about whom I am nominating either, but I am happy to talk about who I think deserves the Nobel Prize among Hungarian researchers. I am thinking of Árpád Furka, who started a completely new method in organic chemical synthesis, a pioneer of combinatorial chemistry, and a great creator. He is recognized in many places in the world, but in Hungary, he has not yet received the recognition he deserves and has not been elected a member of the Hungarian Academy of Sciences. He is one of the few people who have enriched science with truly original ideas and original ideas that have been put into practice.

Who will receive the Ig Nobel Prize?

The Ig Nobel Prize is a satiric prize, it was created as a parody of the Nobel Prize. It recognizes research that should not have been done, and that was unnecessary. But over time, it has also been associated with a certain recognition, because it means that the work has been noticed and this distinction is seldom rejected. Not a shame award, but a humorous award. Sometimes you get the Ig Nobel Prize, and then a few years later the real Nobel Prize, because it turns out that the research that originally seemed superfluous is of enormous value.

What is what you have called a blind spot in the science history of Hungary?

There were many scientists among the victims of the Holocaust, although very few members of the Science Academy, only because of its discriminatory elections. Between the two world wars, the Hungarian Academy of Sciences was anti-Semitic in the spirit of the Horthy regime. The Academy has not yet taken stock of the loss of the Holocaust to academia, although it is important to face up to what happened. Even Albert Szent-Györgyi, who was intimately familiar with the workings of the Hungarian Academy of Sciences between the two world wars, had bitter words of criticism for the way it operated. Even today there are historians who add to the silence and cover-up, and that is a shame on all of us.

We have featured several of your books on our radio station, including books on scientific memorials of Budapest, New York, and Moscow. What motivates you in your writings?

Let me quote someone who wrote a foreword for our latest book. The Nobel laureate Paul Nurse, former President of the Royal Society and now head of one of the world's leading biomedical research centers, wrote it for our book on London's science memorials. When he walks around a city, he is interested in two things: its landmarks and how science is present in that city. He articulated our motivation — the plural refers to the fact that these books are a joint work with my wife Magdolna Hargittai. So, the motivation is, on the one hand, the love of science, interest, curiosity, and, on the other hand, the love of the cities we are writing about. Nurse also writes that although he is a Londoner, he found a lot of new things in our book. He still walks the streets of his hometown with his eyes open and with great curiosity.

For me, it is very important that you combine science and art in your books. You explain concepts so clearly that even a non-scientist can understand them.

In our interest in symmetry, we did not distinguish between scientific examples and artistic examples. These two areas of human activity appeared to us in unison. My second comment takes me back to Wigner. He formulated for me that if we want to explain something to someone, the first requirement is that we ourselves must first understand what we are talking about. When we disseminate knowledge, when we promote science, we are doing a service, but we are also meeting the challenge ourselves that we must first understand what we are talking about before we talk about it. This sounds very simple, but it is not self-evident. Even in high school, I knew exactly when the teacher was explaining something he probably didn't understand. So, the first task is to understand what we want to explain to others. Our participation in the explanation also helps us because it shows whether we have really understood the concept that we have to explain in a way that others can understand.

18

JÚLIA VÁRADI, KLUB RADIO.
JUNE 16, 2021.
Used with permission from Júlia Váradi

You keep writing all the time?

During the Pandemic, my wife and I just wrote and read, and it helped having a good time despite the gloomy circumstances.

How does the collaboration work?

We both write and we both read what the other has written. We're very critical about that, but we don't prescribe what the other writes about or how the other writes it. We don't give up our autonomy. When I was still in the early days of writing, I co-authored a chemistry book with a Canadian English scientist, which was very successful. From my partner, who was the senior author, I learned a lot about what a good textbook should be. But I also decided that I would no longer co-author a book with anyone else (family members being the exception) because I don't like to depend on others. Writing a book is a creation, and only a child can be holier than that. Creating a book between two people is also an intimate relationship and I felt I could never do it again with anyone else. My wife and I work very well together and in the last few years, we have developed this very well with our son.

With Júlia Váradi at the conference hall of the Library of the Hungarian Academy of Sciences following my presentation in the framework of the series "Portraits of Academicians," 2016, photograph by and courtesy of Klára Láng.

The book I have in my hand is about tête-à-tête with science, it was compiled by your son, but it contains your writings. Will the children carry on tête-à-tête with science that you started? Your wife is a chemist, too.

Our son is a professor of chemistry in the United States, our daughter is a professor of the sociology of the Internet in Switzerland and outstrips us all in terms of publications and citations.

So, the name Hargittai sounds good internationally.

Yes, but to carry the name on, that sort of thing, leaves me cold.

So do prizes and awards as I understood from a comment earlier. Does this apply also to academy memberships?

Being a professor, and being the member of a national academy means a lot to me, but that doesn't mean that I don't look at things critically, either at the university or at the Hungarian Academy of Sciences.

*How do you see the future of this mutilated Academy after the govern-
ment has taken away its network of research institutions?*

This is more complicated than how we have seen it recently. Two
years before the changes, I wrote an article about the changes that
had taken place in the Russian Academy of Sciences. I was saddened
to see that the control of science had slipped out of the hands of scientists and that, compared to how important science was in the Soviet
Union, science was much less important in Russia today. I sounded
the alarm bell; I thought that what was happening in Moscow could
happen in Budapest. I had originally wanted to publish my article in
Élet és Irodalom, my favorite literary weekly (Life and Literature), but
its editor thought that it was not an interesting subject. I finally published it in *Magyar Tudomány* (Hungarian Science), the monthly
journal of the Hungarian Academy of Sciences, but it went unnoticed.
Then, it all did happen.

When I referred to the complexity of the issue, I thought that
there could have been a meaningful debate on how to better organize
the network of institutes, whether there was a need for separate
research institutes in each field, independent of universities. However,
the moment it became a political issue, I also became silent. It would
have been impossible to speak as a scientist because everything would
have sounded political.

*Your book just appeared of your conversations with Géza Komoróczy.[1] It
is a biography rather than an interview. As a journalist, and an interviewer, I wondered while reading this huge and extremely fascinating
book, what is your method when you talk to these great people?*

One of my favorite subjects is the art of interviewing. I made the first
one in 1965, at the request of Géza Simonffy, for Radio Budapest
with a Nobel-Prize-winning scientist, Nikolai Semenov. I thought at
the time that I could do interviews, but then of course I realized that
I couldn't. But the first interview went very well because Semenov

---

[1] Géza Komoróczy (b. 1937), Professor Emeritus of Assyriology and Judaica, Eötvös
University, Budapest.

was a skilled interviewee. The interview won an award, was broadcast several times and the text was published several times. Then I stopped doing interviews and for 30 years I focused only on my research. When, 35 years after the Semenov interview, I was compiling my interviews from the 1990s for publication by a London publisher, I remembered the Semenov interview. I bought a copy on tape from the Radio, as it was in Russian, I translated it into English and published it with the others. A few years later, the first two volumes of the six-volume interview series were published in Russian. My English translation of the Semenov interview was translated back into Russian. The interview translated from Russian into English and from that into Russian was, to my great surprise, perfect. The translator was extremely intelligent and Frigyes Karinthy[2] would have had nothing to write about in this case.

In the interviews, you ask questions, and you don't always include them in the published material, because in Komoróczy's case, it's as if he's just talking to himself?

One of the reviews of the Komoróczy book was written by Gábor Miklós and he wrote about the self-restrain of the interviewer. Yes. I learned quite early that the interviewer should not put himself in the foreground. To the question of whether there were any questions left out of the book, every subtitle in the book was also a question.

I suspected that was the case.

There are some topics where it is important that the question takes the form of a question, but in many other cases it is not my question that is important, but what Komoróczy has to say. Can I say something else about interviewing?

---

[2]Frigyes Karinthy (1887–1938), Hungarian writer who wrote a biting parody of how translation may perversely change the original meaning.

Sure, I'm happy to learn.

Well, what you are saying leaves me almost speechless. So, I've developed the principle of medium preparedness. If I'm so prepared that I know everything about the interviewee, I'm no longer sufficiently curious. If I know nothing about the interviewee, that's bad too. The ideal is when I know a lot but retain my natural curiosity. Of course, all I can say in this regard is the amateur interviewer's insights. You obviously know much more about it. I have also learned that for different personalities I need to be prepared differently. For example, I knew in advance about Jim Watson, the co-discoverer of the double helix, that he was a very taciturn person. I could have run out of questions after the first few minutes. In the case of Géza Komoróczy, however, it sufficed for me to be a catalyst. I did not want to know everything about Géza Komoróczy, but about the phenomenon he represents. This a unique and special phenomenon, one that is instructive for everyone.

An institution.

Exactly. And I think I succeeded.

It was an absolute success. I'm going back to the book about tête-à-tête with science. Did your son's selection meet with your approval, or was it discussed beforehand and was the selection a joint work?

This is the second such volume. The first was from my English publications, where the task was much more difficult because he had to select from many more articles. The selection in both volumes was from among my non-technical papers, not from my strictly scientific, professional publications and there were many more published in English than in Hungarian. This second volume was easier.

Favorite subjects?

Symmetry, the history of science, the Nobel Prize, discussions, popularization, and most importantly: the nature of scientific discovery. These run through my writings. At one point, Balazs notes that I write about the same things all the time. It's a little different, it evolves in time, but it's always me.

This is very true. When I was preparing now, but I tried not to over-prepare, I also saw that you were most interested in how a scientific achievement is created and when it becomes a success. In a lot of the conversations you have had with Nobel laureates, you are not only curious about what it was that the Nobel laureate has been awarded for, but also about his personality, his background, what led him to work with such persistence, such curiosity, such unconditional dedication.

I also talked to many non-Nobel laureates of the same caliber as the Nobel laureates. At least ten of them were awarded the Nobel Prize months or even years after the interview.

Maybe you had something to do with that?

That's impossible, but it's not impossible that I helped enhancing their visibility. The magazine, *The Chemical Intelligencer*, which I founded and edited for six years, published many of the interviews originally. It may have played a role in attracting and reinforcing attention. There was a Nobel Prize committee member on the editorial board, and I know that the Nobel Committee for Chemistry was a subscriber to the magazine.

My sense from your writings is that to make it as a scientist you must live abroad rather than in Hungary. The tailwind in Hungary is at best a breeze ...

...or headwinds. Árpád Furka's example always comes up among such considerations. He pioneered a new field, combinatorial chemistry. His method can create entire libraries of peptide molecules, whereas conventional methods take a long time to produce a single peptide. He did not even manage to become a member of the Hungarian Academy of Sciences.

Do you live here in Hungary?

Yes. The great Hungarian American Nobel laureate chemist George A. Olah has a sad comment. He was warned early on when he was still living in Hungary, that if he became even only a little successful, he would immediately be the envy of the world. He encountered this at home, but not in North America. There is no exact English equivalent for the Hungarian word "káröröm," Schadenfreude [malicious joy] in German, and this German word is used in English when people write about such things. Balazs and I are working on a new book about the great Hungarian émigré scientists. We have selected a sample of 50.

Will they be biographies, or will they show why they had to leave to become successful researchers?

The motivation for the book is to show the difference between the two societies. One excludes, pushes talent away, and stifles free speech and free thought. This is the way it is in Hungary, system by system, not system-specific in the history of the twentieth and now the twenty-first century. A non-inclusive, non-tolerant system that lets out, even sends away, in extreme cases destroys — in the case of the persecution of the Jews, literally murders — talent. The other society is inclusive, tolerant, welcoming, giving opportunities, not discriminating. There is a huge difference. The book is written in English, we don't know if there will be a Hungarian version. If it were to be published in Hungary, it might help people to realize how self-destructive the intolerant, non-inclusive, envious attitude is.

Both your children live abroad. Did you ever think of emigrating yourself?

We did. When we were first in the United States, we consciously asked ourselves that question because we didn't want to reproach ourselves afterward for not having thought about it. That was in 1969. At home, I started a new field of study in 1965, we got married in 1967, and in 1970 we had our first child. I am wary of sounding too pathetic in what I am trying to say, but we wanted to assert ourselves at home. I was deported in '44, and they wouldn't want me go to high school in '55, to university in '59, but by '69 things finally seemed to be working out and we could do what we loved. And we did.

Absolutely. But no regrets?

We didn't regret it because we had such a unique trajectory, including our books, that I don't think we could have done the same in America, for example, because we would have been busy building our academic existence. We have no regrets but I will add two things to that. One is that we have had to go through very difficult journeys in our lives to get where we are. The other is that we would be very sad if our children were not where they are now. I should also add that I spent that one year in Texas — my wife had permission to come out only for three months. As much as I loved and still love the United States — now with a little more trepidation and reservation than before — Texas was not the place we would have said was the best destination of our dreams.

But there's Los Angeles, New York.

But we weren't there, and we were directly affected by where we were, so we weren't broad-minded enough. We were drawn back to Budapest, to our parents, to the challenge of the task of creating something new. Was there masochism in that? And romance, of course? We didn't have a home yet but then the hope flickered of buying our first apartment, a room and a half, a kitchen, a bathroom, a small home of our own, with family loans. Where in America could we have found such a place?

I see. Your books include two about yourself, but they are also full of other people. One is Our Lives *and the other is* Mosaic of a Scientific Life. *A lot of them are also based on interviews with your own life experiences, about your childhood, and the horrors you went through in your early years. Can I ask you about that?*

Of course.

Holocaust survivor.

I don't use that term to describe myself. I tolerate it now, but I could not have imagined it before, because I always thought that many people had a much harder fate.

But still, you were in a ghetto, in a camp, and your father died.

Was murdered.

Murdered.

I used to correct myself, too. We tend to dull everything down.

Hungarian custom.

In the Komoróczy book, one segment deals with the semantics trap. My approach has also changed. I grew up thinking how lucky I was to come back from deportation. And today I deeply resent that we were deported. If I try to find an explanation for the attitude that I considered us lucky, I find it in my mother. I know from fellow deportees how she kept their spirit alive, her optimistic attitude under all circumstances. She probably saved lives.

I grew up in that atmosphere of optimism and in that sense, I was lucky indeed. It was better to grow up that way than in anger against the world. But I can see today how terrible it was, I was not even 3 years old when they took me away. Where to? To certain death, because what really happened was only revealed after many years. The destination of our train of cattle cars was Auschwitz. Then, by a strange coincidence of events, our train ended up in Austria instead of Auschwitz. A train that had originally been set up by agreement between Hungarian Jewish leaders and the Germans to go to Austria was mistakenly sent to Auschwitz. Another train was needed immediately because the agreement stipulated that labor would be provided in and around Vienna. This replacement train became our train. The camp was horrible, the *Lagerführer*'s [the camp commander's] cruelties, the starvation, my grandmother died, but there was a chance of survival. In Auschwitz, I would never have had a chance. Many years later I asked my mother what she feared most during the deportation.

The psychological trauma. That's why she thought it was important to always discuss everything, to wash properly, even though it was only cold water, to maintain normality as much as possible. Of course, we didn't completely avoid the psychological damage. That's what I was told because I don't remember that, but I didn't speak for a long time after we came home, I didn't speak, even though I had spoken before.

And your brother?

My then ten-year-old brother, who had been through much worse than me, probably never fully recovered from the effects of the deportation.

Why did you become a chemist? Why are you laughing?

Because in a conversation like this, you can only answer honestly. Chemistry was not my first choice, and that's why I laughed. My first choice was to be a lawyer, obviously, because my father was a lawyer. The only thing I have left after my father is a book he co-authored that I never read. It was a compilation of laws about unfair competition. This book became a symbol for me, from which I derive my respect for books, which may be an interpretation from hindsight. In the 1950s, when I started thinking about what I was going to be, becoming a lawyer was not a real option. The other thing that appealed to me was journalism, in general, the media. I was making films from a very young age. It wasn't real, there was no tape recorder, there was no film. What we had was a roll of paper to put in the cash register. I drew one scene after another on it and then you could draw the paper tape along from the beginning to the end, and whoever sat in my cinema for 20 fillers[3] could watch the scenes frame by frame as I drew the paper tape along. I also understood very quickly about journalism and the media that there was no such thing as real journalism at that time. You had to write about what you were told to write

---

[3]Twenty fillers could be compared to 50 fillers, the price of one scoop of ice cream at the time.

about. The sense of independence and the need for independence that I have had for as long as I can remember has sometimes got me into trouble. My favorite subject was math, I won a local competition in the fifth grade and the prize was a chemistry book, my first encounter with chemistry and I wanted to be a chemist ever since. The book by Péter Teknős, *Treasure of Thousand Colors*, was about coal, and what you could make out of it, far beyond the fact that you could burn it to make heat; it was very modern in that sense.

What is chemical symmetry? Why did you bother?

My interest in symmetry had two sources. One was photography. I photographed the construction site in Újpest, visible from Margaret Island, and it was reflected in the Danube River. It was almost a perfect mirror image. Then a breeze came and the next picture, although still recognizable as a reflection, is far from perfect. This was a fundamental question for me, how much can a reflection be spoiled so that it is no longer seen as a reflection. In nature, symmetry is not perfect, only in geometry, and it is not only interesting but valuable for the acquisition of knowledge. The other source was my meeting with Eugene P. Wigner, who in 1969 personally introduced me to scientific knowledge about symmetry. This was at the University of Texas, Department of Physics, where I was a visiting research associate for one year and Wigner came for a short visit. We had already known each other through correspondence because in 1964 my first published article was a response to a philosophical paper by Wigner. Both were published in *Élet és Irodalom* (Life and Literature). There is no specific chemical symmetry, only symmetry has very important appearances in chemistry and my wife and me published a book on this in 1986. It was a great success; the third edition was published in 2009. There are universities where it is used as a textbook or as a reference book. The most beautiful and best-known chemical manifestations of symmetry are molecules and crystals, but it is not so obvious that certain chemical reactions take place or even that they cannot take place as determined by symmetry properties.

You were associated with many great scientists, Watson, Wigner, Teller, and many others. What I'm particularly interested in is, did these great scientists have anything in common?

First, they understood what they were doing, which is not always self-evident. And they could explain it, and make others understand it, and not everyone can do that, only those who really know what they are doing. It is a great experience when, for example, Edward Teller explains the uncertainty principle. I wrote a big biography of Teller. I have learned so much about his life that it is almost indiscreet. Sometimes there is a family argument about something and then Wendy Teller, Teller's daughter, asks me what the truth is. The reason I have dealt with Teller's life in so much detail is because I had previously dealt with the lives of the so-called Martians. My most successful book to date was about the five Martians. I wrote this book on request. I had previously thought that everything had been written about them, and Oxford University Press, where I had already published a well-received book about the Nobel Prize, asked me to write a book about these five Hungarian scientists. They are the ones who even risked their scientific careers to do something more important at the time, the defense of the United States and the Free World. The five Martians all came from upper-middle-class Jewish families in Budapest. George Marx called everyone who had emigrated from Hungary to the West and promoted Hungary's reputation there, a Martian, a euphemistic masking of the facts. He did not want to write about their Jewishness.

That became the adjective instead of Jew.

In Hungary, it is acceptable to write that the communist leader Béla Kun was a Jew, it is acceptable to write that the ruthless dictator Mátyás Rákosi was a Jew, but it is not acceptable to write or talk about Wigner, von Neumann, Teller, von Hevesy, or Olah. It is all part of the failure to face the past. Unfortunately, the Hungarian Academy of Sciences can be condemned for this, as can the entire Hungarian officialdom. The mention of their Jewishness is not diminishing their being Hungarian at the same time. In many respects, they became what they have from the Hungarian cultural environment, from the

Hungarian system of high schools, but their Jewishness, their family background, their ambitions, and everything else that has hindered their domestic success because of their origins, are part of the whole picture.

I imagine that in the book you and your son are currently writing on Hungarian émigré scholars, you will emphasize this.

Of course. We want to contribute to a more complete picture, to put things in their proper perspective.

When you think back on the conversations you've had with extraordinary people in your life, is there anyone you think of as the most extraordinary, that if you had missed, your life would have been poorer?

The long conversation with Teller, the correspondence that followed, and the work that resulted in a 550-page biography, combine to make this a special experience. If we look only at the conversations, I will highlight the one with a Harvard professor. Frank Westheimer told me how his studies had developed, and how he would have liked to work alongside the famous professor James Conant. By the time he got there, Conant had become president of Harvard and couldn't be his mentor. When Westheimer finished his studies, he went to see Conant for a farewell visit, and Conant asked him about his plans. Westheimer told him, and Conant replied that if Westheimer succeeded in all of this, it would still be a footnote to a footnote to a footnote in the history of chemistry. Then they said goodbye. Westheimer thought deeply about Conant's words. He hadn't thought until then how important everything he was doing could be. But from then on, throughout his career, he had weighed everything against what Conant would say. Conant's words have haunted him all his life. Westheimer was in his eighties but continued his research as a retired professor at Harvard University. One day, someone knocked on his door. It was Conant, and he asked Westheimer if he remembered him. At this point, Westheimer became so emotional that we were unable to continue the conversation, which ended there, and I am emotional when I tell this story. Often a warning, a piece of advice, or a question can become a life-changing experience. This also happened to me. Westheimer's story is so human and so much not special that it is so important that he told it, that he put it into words, that he captured it. It is worth reflecting on whether there was such a powerful event in our own lives that we might have to mine out of the many memories, because when it happened at the time we might not have noticed it, but the impact has stayed with us. In this case, the story is not so much about Conant's greatness, but about Westheimer's greatness, because he had the ability to transform a possibly innocuous remark into a life motif. Ágnes Heller wrote a beautiful foreword to my book *Mosaic of a Scientific Life*, one of her last, perhaps her last, writings.

She wrote a beautiful Foreword.

Agnes Heller presenting the Hargittais' New York book at the conference hall of the Library of the Hungarian Academy of Sciences, 2017, photograph by and courtesy of Klára Láng.

It says that excellent people often have excellent teachers. But it may not be that those teachers were so much more remarkable than others; rather, it may be that the excellent student had the ability to extract extraordinary messages from the teachings and experiences.

He could be well taught.

Exactly.

* * * * *

Csaba Károlyi, "Something about the Art of Interviewing." Élet és Irodalom, June 25, 2021 (literary weekly). This is a review of the preceding interview. Used with permission from Csaba Károlyi.

Last week, Júlia Váradi talked to academician István Hargittai in Klub Radio. Hargittai works on structural chemistry and publishes in many languages, but the general public at home and abroad is familiar with his interviews with scientists and his biographical books. He was Váradi's guest for the publication of his conversations with Géza Komoróczy entitled *In the Footsteps of Jeremiah*. This book is an outlier in Hargittai's career, as he has so far presented natural scientists. His most famous book is *The Martians of Science*, which explores the work and personalities of Theodore von Kármán, Leo Szilard, Eugene P. Wigner, John von Neumann, and Edward Teller.

Géza Komoróczy is an orientalist, historian, Hebraist, and Assyriologist, who has studied Sumerian literature, the Hebrew Bible, and the history of the Jews in Hungary. Why did Hargittai choose Komoróczy as the 'subject' of his new book of seven hundred pages? That was the question that came to me. Unfortunately, the conversation was not very much about this. What I got instead was also interesting.

Júlia Váradi first asked about her subject's family. His children are scientists working abroad, but his wife and son are also involved in the work on the books. Asked about the mutilated Hungarian Academy of Sciences, Hargittai also mentioned *Élet és Irodalom*. He had written an article about the critical situation of the Russian Academy of Sciences, which foreshadowed the troubles of the Hungarian Academy at a time when it was not being paid attention to at home, and he recalled receiving a reply from the magazine that it was not an interesting topic. When they came to the Komoróczy volume, I began to rejoice. "It's an interview book, but not an interview," Váradi said. You don't ask questions? Or do you skip your questions? I ask a lot of questions..." Indeed, sometimes the questions in the book are: "Your relationship with Imre Kertész?" — Hargittai replied that in most places his questions were not the important ones, and the subtitles were all questions. Váradi, as so often, immediately interjected. "You almost left me speechless ... Can I say something else about interviewing?" — asked the poor recent interviewee. "Medium knowledge is good; you shouldn't over-prepare. And you must prepare differently for everyone..." Now, I would have listened to that for a long time, but then unfortunately the interviewer changed the subject.

The topics of the conversation included the nature of scientific success, why Hungarian scientists go abroad ("it's not that they go, it's that they leave for good"), why chemical symmetry is interesting, which is the author's most extraordinary conversation. "Why didn't they emigrate?" — was a good question. "In 1969, in America, it came up," came the good answer, "but we were drawn back to Budapest, we had a hope to acquire a small apartment of our own, and where could we find that in the United States?"

In the end, four recent books came up, walks around the scientific memorials in four cities, London, New York, Moscow, and Budapest, published also in English and Russian. (Hargittai publishes only a selection of his books in Hungarian. In the Komoróczy volume, his dazzling English-language books are not mentioned, he directs us to his website.)

Towards the middle of the rich discussion, we only found out why the book, *In the Footsteps of Jeremiah*, was made. Hargittai is a Holocaust survivor. He did not use that term to describe himself, he noted, and pointed out that a sub-chapter of the Komoróczy book was about the "semantics trap." Why do we use euphemistic terms such as labor service, and the Holocaust? Why is a person of Jewish origin reluctant to reveal that he is Jewish? Hargittai and his family were deported, coincidentally not to Auschwitz but to Austria. He felt lucky to have come back, but today he also says what an outrage it was to have been deported. (Hargittai asks Komoróczy in the sub-chapter: "do you see any hope that one day this camouflage will cease, and the language will be cleared up?") Somewhere here can one find the reason why the idea of these conversations occurred to the professor of chemistry. And in the fact that Komoróczy hit the wall of the false myth of "national greatness." As the author puts it in a sentence that appears on the back cover of the book, "What is it if not going against the wall [as the prophet Jeremiah, Komoróczy's true prophetic persona did] when the scion of a noble family becomes of Jewish identity in a largely anti-Semitic environment?"

Anyone who wants to know more about the unique and fascinating personality of Géza Komoróczy should read the book. If you want to know the very consistent personality of Istvan Hargittai, hidden in his interviews, read his interviews. One more little thing. Júlia Váradi asked him why he became a chemist, and Hargittai laughed. "Why are you laughing?" — came the question. "Because in a conversation like this, you can only answer honestly," he said, and then answered honestly. But perhaps he laughed because he saw himself in Váradi. As a questioner.

19

CONVERSATION WITH LÁSZLÓ NYULÁSZI AT THE BUDAPEST UNIVERSITY OF TECHNOLOGY AND ECONOMICS, NOVEMBER 28, 2021, CELEBRATING ISTVAN HARGITTAI'S (B. AUGUST 11, 1941) 80TH BIRTHDAY, IN FRONT OF A LIVE AUDIENCE.

Used with permission from László Nyulászi

Let's start with some biographical information.

I was born in Budapest in an "assimilated" Jewish family. In 1942, my father, a successful lawyer, was taken away to a slave labor camp, he was killed, and we moved to Orosháza in southeastern Hungary, where my mother's family lived. In 1944, we were deported. I was less than three years old when the train of cattle cars left for Auschwitz. Another train carrying deportees to Austria had mistakenly gone to Auschwitz and a train to Austria was needed quickly. This became "our train." The deportation and the whole Jewish background indirectly prepared me for later trials. One of the "achievements" of the 1956 revolution was that I could study in the high school of our town where in 1955 I had been denied admittance on account of my late grandfather's having been a store owner. So, I started my high school studies in Budapest. In 1957, I continued in Orosháza without ever having to discuss with the people in charge there, what had happened. There was one incidence though. Not knowing the local rules, I went up the girls' staircase. This was at the most critical time when some of the students at this high school were participating in the revolution and its aftermath, but the janitor, whose tasks included guarding the rules, found it important to report on my infringement of the rules to the principal of the school.

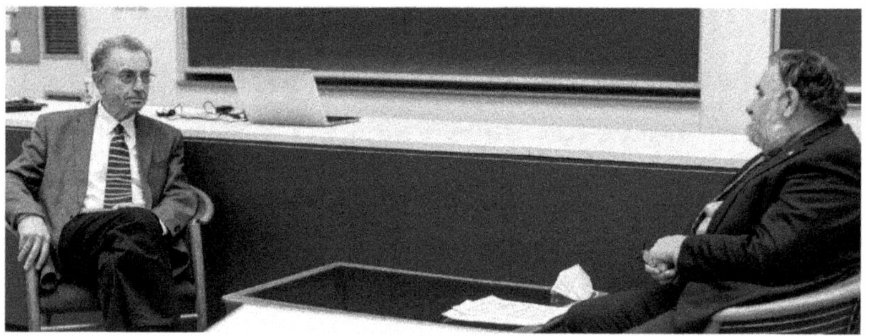

With László Nyulászi (right) at the Grand Lecture Hall.

Did you have any experiences at school that later proved to be formative?

I had brilliant teachers, but all the time I was wondering if I would be able to go on to university. I found it difficult to get into high school and then, university.

You studied in Moscow and how did you choose chemistry?

I'm glad we're taking a big leap. Mathematics was my favorite subject. Back in the fifth grade when I was eleven, I won a small math competition and got a chemistry book as a prize. This chemistry book was full of politics, but I didn't notice that at the time, I filtered out what I was interested in and only remembered, for example, what could be made from coal. In this respect, this book from around 1950 would be still up to date today, because it emphasized that the least remarkable application of coal was burning it. I started experimenting at home, doing dangerous experiments. I was at home alone in the afternoons; I could buy everything they sold at the "household shop." Later I didn't become a chemist experimenting with chemicals, but I enjoyed it then. Overcoming some hurdles, I got into Eötvös University to study chemistry and completed two years. My love of mathematics and physics remained, but you don't get much of that for chemistry majors, and the curricula at the university were rigorously structured and compartmentalized. One of the attractions of going to the Soviet Union was that chemistry students there could study a lot more math and physics. That year the only option in chemistry was radiochemistry but I signed up for it.

Was it advertised? How did you learn about it?

One of my former classmates told me about it. He went straight to Moscow after graduating from high school. There was another attraction to studying abroad. I had previously been categorized as a "class alien," the most disadvantageous social category, which almost prevented my acceptance to university. I had to pay tuition (that is, my parents did), whereas going to Moscow meant that I would not have to pay tuition, rather, I would receive a scholarship. So, I applied for it, and it was not difficult to get in.

I was supposed to go on to Leningrad to study radiochemistry at a small university that specifically trained radiochemists, but I didn't want to be a radiochemist. Lomonosov University in Moscow, however, appealed to me. I don't even vaguely remember how I moved from one group to another, but nobody noticed, there was no high level of organization, so I obviously took advantage of that. It was not a long thought, it was more an impulse, a sudden decision on my part. It had, though, important consequences. As a radiochemist, upon graduation, I was supposed to go to work at the Central Research Institute of Physics. However, since I would not graduate in radiochemistry, that job disappeared. When I understood this in my third year, I started looking for a specialization that I would like best. Lomonosov University was rich in research opportunities. In my third year I went through all the laboratories that I thought would be interesting. The Department of Chemistry at the Lomonosov is also a huge research institute, and it took me a year of doing this before I chose molecular structure research using electron diffraction. I was attracted by the discipline, I was attracted by the people, and I was also attracted by the fact that they had very good international contacts, which was quite rare in the early 1960s, not only in Moscow but also in Hungary.

For me, it is new that you could find a research group there that was internationally renowned.

It wasn't very common, but there were some. The staff at this laboratory had an international outlook. They were not given any help to do so, but they were not particularly hindered either. It may even be that this laboratory was not so significant as to be conspicuous. When I was appointed a university professor back in Budapest in 1990, which was after the political changes, the Secretary of State held a reception and had a talk with me on that occasion. He asked me to compare my experience in Moscow with my long-time experience in the United States, in terms of laboratory activities and international cooperation. He phrased his question in such a way as expecting to hear about the constraints in Moscow. In America, I enjoyed great freedom and opportunities but my personal experiences in Moscow were also positive, even if staying informed of the international literature required considerable extra effort. In Moscow, my colleagues were ready to make such an extra effort, and they inculcated in me the importance of international interactions. There was sizzling scientific life in the laboratory with passionate discussions from which I learned a great deal. I would also like to share the atmosphere in which this discussion with the Secretary of State took place. At the time, around 1990, it seemed that earlier Soviet connections were not worth boasting about, and many people preferred to keep quiet about them. One of my academician colleagues who is still around, declared that he did not speak Russian whereas I knew that he had also studied in Moscow. In the Moscow laboratory where I did my thesis, the first requirement was a detailed and up-to-date knowledge of the international literature. This was not trivial at all. Not very many years before, the application of the resonance theory in chemistry was declared a product of alien (that is, Western) ideology, and eminent chemists who had applied it, lost their jobs. Even at the time of my studies, there were similar "forbidden" areas of knowledge though their uses did not have so harsh consequences. When I was graduating, my Russian professor suggested to continue there for my PhD-equivalent degree, and I would have been happy to stay.

During the conversation, turning toward the audience.

This offer was the most attractive because I did not have a job back home. In Budapest, I visited the Secretary General of the Academy of Sciences to get the necessary permission. The Secretary General asked me if we needed electron diffraction molecular structure research, and I said yes. Then, he told me, come home and implement it. At the time I felt this a rude intervention, but today I am grateful for it. This decision immediately put me on an independent path, and forced me to look for a job where I would have the opportunity for independent development. I found it in a recently organized small research laboratory of the Science Academy located at one of the university halls. There, I developed my research literally from scratch and without any special financial support. I acquired instruments and built a group from colleagues who were not very successful in other research groups and were eager to do something promising.

Right at the start, I contacted four foreign research groups for unpublished documentation that could help me build my experiments. Such work is not primarily aided by published papers, but rather by papers that describe the trials, possible dead ends, and the small details of the experiments. I approached two Americans, one Japanese, and one Norwegian. One of the Americans did not reply, the other wrote about why he thought my situation was hopeless. The Japanese sent me four Japanese-language internal reports, which we had translated, and which proved useful. The Norwegian invited me for three months to study their work on the spot and offered me all the help I needed for such a visit. It was Otto Bastiansen, it was to become a lifetime friendship and a long-term, most fruitful cooperation with his associates.

There was a particularly fortunate circumstance in the small academic laboratory in Budapest. There was a well-equipped mechanical workshop with a manager of excellent ability, József Hernádi. He had four staff members. It was a special opportunity, rare even then, almost unthinkable today, to build new experiments. In addition, Hernádi was happy to take up the challenge I brought him because until then he had to fulfill rather pointless orders. It was a matter of prestige how much workshop time was used by the various members of the otherwise small staff of the laboratory, but there was hardly any meaningful demand. Everyone preferred to buy ready-made instruments, and I was looking for genuinely new solutions by combining my own ideas with those of Hernádi. A few years later, József Hernádi told me that in his favorite novel [a book by Lajos Szilvási], a young engineer arrives at the plant full of ideas, the others tend to block him, and then he manages to realize his ideas with the help of an old workshop foreman. We built experiments that broke new ground in our field of science. Soon, the American professor who had given me the discouraging answer to my inquiry asked us to make a replica of one of our devices, and he came to Budapest personally to pick it up. There was no mechanism for accepting money, so he reciprocated with a sophisticated calculator that meant a lot to us at the time.

And you also got to America, which was another special opportunity at the time.

During the visit to Norway, I met Harold Hanson, the head of the Department of Physics, University of Texas at Austin. His university invited Otto Bastiansen for a visiting professorship. Bastiansen was not only a distinguished scientist but also a Rector of the University of Oslo and later President of the Norwegian Academy of Science and Letters. He received an elegant invitation from Texas, which included two stipends each for a one year stay for young associates of Bastiansen's choice. I was to be one of the two. It was difficult to receive permission for this visit. I was supposed to start on September 1, 1968, but it was not until the end of January 1969 that I was able to join Bastiansen and his Norwegian associate in Austin. The year in Texas was a defining period. I had the opportunity to meet greats like Eugene P. Wigner, from whom I learned about symmetry, and Michael Polanyi. Opportunities appeared almost like a chain reaction.

But you went after them, too. Your meeting with Wigner was not by accident.

I was at the Department of Physics and saw the announcement about Wigner's forthcoming lectures. I had been in contact with him since 1964. Back in Hungary, he was known as an "expatriate son of the fatherland" but was only accepted after his Nobel Prize in 1963. In 1964, an article about Wigner was published in the literary magazine *Élet és Irodalom*. Wigner was so moved by this that he sent a philosophical article to the magazine for possible reprinting. The article was about the frontiers of science. *Élet és Irodalom* translated it and published it with a comment that Wigner would be interested to know what people in Hungary thought about the frontiers of science. I was a university student in Moscow, and the only Hungarian paper I was reading was *Élet és Irodalom*. I saw the article and responded. Looking back, I am surprised because I had never published anything before. So, I wrote a response and sent it to the paper, which ran it. To my even greater surprise, a few weeks later I received a beautiful letter from Wigner and a bunch of reprints of his papers. Thus began our interactions. In 1969, I indicated to the chairman of the Department of Physics (by then Hanson was elsewhere) that I would like to meet Wigner. The chairman said that this was out of the question. Wigner was coming for one week and they would use every minute of his expensive visit. I left a little note for Wigner saying that I was there, where my office was, and where my laboratory was. The morning after his arrival he knocked on the door, introduced himself, and from then on, he came every day during his stay at 8 o'clock and talked to me about symmetry until 8.55 when he left to appear on time for his official program, which started at 9 o'clock. He used his private time on me, at no loss to the Department. At the time I almost took this gesture as something natural; now I know what a special gift it was.

I teach general chemistry, right in the first semester, and employ a very simple, rudimentary theory whose beauty is in its simplicity and broad usefulness.

One of the shortcomings of the Moscow course was that this theory was not taught because it uses the concept of electronegativity, among many others. Electronegativity was frowned upon for silly, ideological reasons in Moscow at the time. In January 1969, on my way to Texas, I stopped for a short visit at Indiana University to give a lecture on the first results of our newly initiated research on the molecular structure of simple, sulfur-containing compounds. During the discussion, someone suggested the possible use of this theory, called Valence Shell Electron Pair Repulsion, VSEPR, in short, theory or model. That night, I went to the library of Indiana University, I was not sleepy because of the jet lag, and wrote a long letter on this theory, illustrated with colorful diagrams, to my wife — we had a short time before married — who was a graduate student at the time. On my return from America, I wrote an article about this model for a Hungarian magazine whose name now is *Természet Világa* (World of Nature). It was science popularization, and I used the simplest molecular structures for illustration. To my chagrin, the model did not seem to work exactly for these simplest examples. I wrote to the creator of the model, Professor Ronald Gillespie in Canada, and he replied immediately, took the problem seriously, but asked me not to rush into publicizing my critique because he had just published a book about the VSEPR model. A great collaboration developed, culminating in the publication of a co-authored book 20 years later, which was republished after another 20 years.

The Gillespie–Hargittai book on the VSEPR model.

I may add that the problem I noticed was solved before we produced our joint book. I could make the model more broadly applicable. We had a great deal of amicable discussions during the preparation of the first edition of the book. I succeeded in convincing Gillespie in some matters, but generally, I learned a lot from him. He taught me that, especially when writing a textbook, every page, every paragraph, must be didactic. You must not leave things vague, and whatever goes into the textbook the author must understand it first. This sounds like a

At the end of the conversation with a diploma of appreciation, a medal, and a bottle of wine.

trivial requirement, alas, reading the scientific literature has taught me that this is by far not always the case. Logistically, it was not an easy project. Our book was essentially written by me, but every page was corrected, critiqued, and tightened by Gillespie, and we didn't accept anything as final until everything was crystal clear and unambiguous. Gillespie had retired at the time of writing the book and was sailing the Intercoastal Waterway of North America. I had to know which port to send the manuscript of the next chapter to. We didn't use e-mail then.

Your work with your wife is an example of a special collaboration, very fruitful. How did this develop?

We started "dating" when she was a student (not my student), a sophomore, and she was 21, close to 22 when we got married (I was 26). She started doing extracurricular research with someone else to avoid the situation where my wife would also be my student. But over time, she was so attracted to my field that she switched. She did her thesis with me, but officially with the director of the laboratory as her thesis advisor. After graduation, she stayed on at our laboratory. We had no problem working together until some of our colleagues noticed that we were very productive. Then I was called in for a discussion in front of the leadership of the laboratory. It consisted of the director, the communist party secretary, the trade union representative, and the personnel officer. I was told that husband and wife could not work together. I replied that I would start immediately looking for a job *for me* elsewhere. The discussion stopped and that was the last time anyone would ever bring up the subject. They assumed that my wife would leave and were dumbfounded by my response. There was really no rule that would have excluded our working together. As for me, I have no idea where I could have found another place in Hungary to build up my research. My response was not thought out, it was on the spur of the moment. Going abroad was not a possibility at that time with hermetically closed borders. As for the laboratory, had I left, the only research direction that corresponded to the goals of this recently established unit of the Academy would have disappeared. My wife and I continued working for many years, until 1990, when I left for the University of Technology and Economics. Thematically though she had become independent long before and became extremely successful in the field she had developed. We have continued working together on books.

Your transfer to the University of Technology was not a usual thing.

In America, it is almost impossible for someone to be appointed professor where they had studied. Hungary is of course too small for that, but the mobility is even less than what would be achievable even under our circumstances. A former dean of Faculty of Chemical and Bioengineering, Sándor Gál, implemented unusual solutions, and those included the professorial appointments of non-homebred associates of this university. Even at the time of economic austerity, he developed conditions for change. I was invited by him to become head of a department, which I accepted for 5 years. Few may have believed that I meant this limit; he did. He understood that I considered myself a professor and a researcher and for me the chairmanship was a duty, a service that others had to undertake just as much as I did. It's not a good thing to get too comfortable in an administrative position, because there is a danger that after a while it becomes difficult to get back to substantive work.

With your books, you have created almost a new genre. I am thinking now of your talks and your works on the history of science, which are far more than mere science history.

What concerns me most is the nature of scientific discovery and, of course, the nature of the scientific explorer. How a scientific discovery is made and who becomes a discoverer. However, I think that perhaps we should end our discussion now, when we have still had the interest of our audience alive, and not when everyone had already been exhausted.

One more thing I would like to ask. You used to have your conversations exclusively with natural scientists, well beyond structural chemistry, chemists, physicists, and biologists.

For me, these conversations meant a second university.

But now I see that you have broadened your interest and opened to social sciences. You have published a book of your conversations with Géza Komoróczy, a famous historian, assyriologist, and scholar of Judaism. Could artists and politicians be the next?

With André Goodfriend (right), fall 2015, Duncansville, Pennsylvania, by Magdolna Hargittai.

No, not at all. I have had two occasions when I stepped out of the realm of the natural sciences. One was Géza Komoróczy and the other André Goodfriend, a U.S. diplomat. He served here in Budapest between 2013 and 2015 as head of the Embassy in the absence of an ambassador. During his tenure, he was unusually active politically. We met a couple of times during his time in Budapest, but our plans for further meetings did not pan out. After his departure, we met in the United States and we both found that we should make a book out of our conversations. Goodfriend came for a few days to visit our son, who is a chemistry professor in Pennsylvania, and we visited Goodfriend, also for a couple of days, in his home in Virginia. We recorded many hours of conversation. At the time, he was still in the U.S. Foreign Service, so he had to ask for and received permission from the State Department to participate in this book project. When we completed the manuscript, he presented it to the State Department. He was told not to publish the book though there was no explanation of the reason for their decision. Goodfriend is due to retire in April 2022, after which the book can be published, as it will no longer need State Department approval.[1] The most interesting feature of our conversation was to follow the evolution of the increasingly autocratic Orbán regime between 2013 and 2015. There's also interesting information about the U.S. Foreign Service and about Goodfriend's career and his diplomatic experience in other posts.

The Komoróczy book is entitled *In the Footsteps of Jeremiah*, published in the spring of 2021 by the well-respected literary publisher Magvető. We recorded our conversations in 2020 amidst the frightening Covid excitement, first in person and then by email. It was a great experience for me to talk with this world-renowned scholar of Assyriology and Hebraism. It was a welcome diversion for both of us in the increasing isolation of the Pandemic, working on this project.

---

[1] André Goodfriend retired from the U.S. Foreign Service in April 2022 and our book has appeared: Istvan Hargittai, *Open Government, Open Diplomacy: Conversations with a Former American Diplomat M. André Goodfriend*. Budapest: Central European University Press, 2023.

20

MARJORIE SENECHAL, "TURNING THE TABLE: A CONVERSATION WITH ISTVÁN HARGITTAI"

The Mathematical Intelligencer 2023 Spring issue, 45(1), 73–77. Reproduced with permission from Marc Strauss, Publishing Director, Mathematics Journals, Springer Nature.

Surely no nonmathematician has done more to highlight the international mathematical community than the renowned chemist Istvan Hargittai. Mathematics, chemistry, physics: his essays portray the international scientific community at the turn of the millennium, in the words — mostly — of the scientists themselves. The portrait created by those essays is notable for its depth and its breadth and for its clarity, insights, and charm. Moreover, in many respects, it has been a family affair. Magdolna Hargittai, Istvan's wife, has been his scientific collaborator since their marriage in 1967; their son, Balazs, also a chemist, joined in later. (Their daughter, Eszter, though not a chemist, shares their passion for excellence in exposition: she is a professor of communications and media *research.) This is how Istvan Hargittai describes his focus on building communities:*

> I always felt a need for building communities, I started my univer-
> sity studies in Budapest and completed them in Moscow. I enjoyed
> the friendship of my Moscow colleagues and decided to ease the
> rigid separation of the two worlds, East and West. I organized
> meetings and invited Soviet scientists to Budapest. Such exchanges
> were not easy even between "brotherly" socialist countries. For
> me, it is painful to see the isolation of Russian scientists once again,
> this time because of the aggression and criminal actions by Russia
> against Ukraine.

Istvan's interviews with Nobel laureates and other eminent scientists fill six volumes. I decided to turn the table, so to speak, and interview him. But this time, the table is virtual: we conducted this interview by e-mail. I'll let his son introduce him. With his permission, these biographical remarks are excerpted (lightly edited) from Balazs Hargittai's collection of Istvan's essays, Culture and Art of Scientific Discoveries *(Springer, 2019).*

Istvan was born in 1941 in Budapest into a happy, "assimilated" Jewish family. His father was a successful lawyer whom the anti-Jewish laws of Hungary made into a slave laborer and who was killed in 1942. Istvan was not three years old yet when the train of cattle carriages started his journey to Auschwitz. By some fatal error, another train carrying a selected group of people had been sent to Auschwitz instead of Austria, so Istvan's train was rerouted to Vienna to replace it. His mother and ten-year-old brother survived the war, though his grandmother did not. After liberation they returned to Istvan's mother's hometown in southeastern Hungary, where he grew up, having a loving stepfather (without having been officially adopted) who had lost his wife and son in another camp in Austria.

Hungary soon became a Soviet-type totalitarian state. Istvan was declared "class-alien," the most disadvantaged social category. Although he excelled in his studies, the local authorities prevented him from entering the local high school on account of the fact that his late grandfather had been a shopkeeper. The higher authorities allowed him to continue his studies in Budapest, but the ordeal was repeated when he faced hurdles in getting admitted studying chemistry at Eötvös Loránd University in Budapest, this time on account of his stepfather, also a former shopkeeper. Istvan obstinately fought to change his "class-alien" label to "intellectual," which he should have had in the first place on account of his late father. He succeeded, but the social categorization was rapidly losing its significance. From 1961, he continued at Lomonosov Moscow State University and graduated in 1965 with the equivalent of a master's degree.

Back in Hungary, Istvan worked for the Hungarian Academy of Sciences and created an area of research, molecular structure determination by gas-phase electron diffraction that was new in his country. He became internationally recognized and earned his PhD-equivalent degree and then his DSc degree, and for quite a period he was a leading figure in his field. Istvan and Magdolna, also a chemist, married in 1967. Since then, they have worked together, done much research independently, and developed projects jointly that they both enjoyed and that added to their fame, like applying the symmetry concept in chemistry and almost everything else, twentieth-century science history, interviewing famous scientists, and writing books.

"My parents are still active," Balazs concludes his remarks. They "still work mostly on joint projects, and still appreciate each other and everything that they are doing — they involve themselves only in projects they can enjoy." I hope that Istvan has enjoyed being interviewed.

Northampton–Budapest, October 2022

What in your background and education led you to, or pointed you toward, the broad, humanistic view of science for which you are renowned today?

For a considerable period of my career, I was narrowly focused. I had to create my experiments and my laboratory, and eventually, build a group. I focused on molecular structures, their determination and modeling, and preferred to read journal articles to monographs, thinking that by the time new knowledge had become part of a monograph it had already lost some of its relevance. This narrow focus lasted from the start of my work in 1965 until about 1990.

What brought you to chemistry?

My interest in chemistry traces back to my having won a local mathematical competition when I was ten years old, for which I received a chemistry book. It was a highly politicized book, almost sheer communist propaganda, which I filtered out at the same time that I learned fascinating things about what can be produced from coal, for example. In hindsight, it had a modern approach, because, for example, it said that it is much more economical to use coal for chemical transformation than to burn it to produce heat. I never lost my interest in mathematics, but I understood that I would be good in some areas only and not at all in many others. My excellent math teacher was uninterested in encouraging extracurricular activities. Later I learned that he was aiming at an academic career — he eventually had a splendid one — but was also disfavored by the regime for advancement.

Would you say a bit more about how disfavoring worked?

In the Kadar era (named for the communist chief János Kádár), which lasted from 1957 to 1989, there were ups and downs for liberalization. It may have looked continuous and gradual toward 1989, but there were moves ahead and backward. From the mid-1960s, for a few years, there was some easing up and then a drastic cutback in the early 1970s. For example, I was given permission to visit the University of Texas at Austin, which had been arranged by Otto Bastiansen, a leading scientist, chair of the Research Council, rector of Oslo University, and later president of the Norwegian Academy of Science and Letters. He was a big-name Norwegian professor, a wonderful person, and a friend of the Soviet Union. He had received a one-year visiting professorship from the University of Texas at Austin and not only with a big stipend: he could also take two junior associates with him, and one of the two was me. But though I was to have departed on September 1, 1968, I wasn't permitted to go until January 1969.

I experienced a setback around 1978 when my brother left Hungary "illegally." I almost became a nonperson. Although I could continue my research, my circumstances changed. That year, I was invited by the Norwegians to give a big lecture in Oslo as part of the celebrations of Otto Bastiansen's 60th birthday. But I was not permitted to go. My complete lecture was translated into Norwegian and published. Some of my friends were afraid to remain my friends, and my superiors at the science academy treated me as if I had been infected by a contagious disease. I was officially informed that because of my brother, I could not receive a passport for foreign travel — a purely Nazi approach, to punish family members for the "crimes" of other family members.

But I was allowed to visit the University of Connecticut under a larger agreement between the Hungarian Academy of Sciences and UConn. I was a visiting professor at the University of Connecticut, Storrs, first of physics in 1983/84, then of chemistry in 1984/85; my wife, Magdi, and our children were also able to accompany me. Compared with previous permissions, this time (1983) it was relatively easy. But our Storrs professorship, too, was preceded by refusals. For example, in 1982, I was invited to give a plenary lecture at the meeting of the European crystallographers in Jerusalem. It was a big affair. There were signs that I might be permitted to go, but in the end, I was not. There was big media coverage of my no-show in Israel. I gave a plenary lecture instead at the next European crystallography meeting, which was in Liège in 1983. We did this on our way to Storrs.

Storrs was a seminal road post in our lives. Our children attended local schools, learned English, and acquired an international outlook. It was in 1984 that you, Marjorie, invited us to attend your "Shaping Space" conference, and we visited Smith College. We took a snapshot of Eszter at the college gate for a souvenir; we could not have dreamed at that time that one day she would attend Smith. (Eszter Hargittai graduated from Smith College in 1996.)

And how did your scientific outlook broaden?

I was left with extra time, and I started writing a small book about symmetry: *Symmetry Through the Eyes of a Chemist*. The much-expanded version of this small Hungarian book later became an international success in English. It included a group-theoretic treatment of various applications of the symmetry concept in chemistry; that was the work of Magdi, who became my coauthor. The third edition appeared in 2009. Jokingly, I said that the Hungarian secret police had harassed me into a career as a writer following my brother's escape.

Your question concerning my humanistic interests made me dig deeper in my memory. Before I knew that chemistry existed at all, I had two professions in mind in quick succession. My first idea was to become a lawyer. That was due to having had a lawyer father, though I did not know him. He coauthored a book about unfair competition, worked in corporate law, and took up pro bono cases. He was the youngest sibling in a large family and the only one to acquire a college education. He worked throughout his studies. But I understood that being a lawyer was not a viable profession in communist Hungary around 1950. In 1942, my father was sent to a Hungarian slave labor camp on the Eastern Front, deep in Russia, where he was forced to search for land mines with his bare hands. He was killed when a mine that he had uncovered exploded.

My next choice was to become an investigative reporter, as we would call it now. I was interested in the media. I had rolls of paper from the cashier's machine and drew the news on it, frame by frame. I let anybody watch the news as I was rolling the paper from the beginning to the end, my "newsreel," for a modest fee of 20 fillers (100 fillers were the equivalent of 1 Hungarian forint and at the time a scoop of ice cream cost 50 fillers. Today (2022) it is 500 Forints.) Perhaps the idea of an independent journalist was even more alien to communist Hungary than being a lawyer.

The political changes in 1989/90 made it possible for me to move from a purely research position to a university appointment when the University of Technology invited me to a chair. This move encouraged me to reignite my broader interests. The example of *The Mathematical*

Intelligencer led me to initiate *The Chemical Intelligencer* in 1994. Springer warmly welcomed my proposal, and a publishing director, Gabriela Radulescu, was instrumental in getting things moving. My interviews became a solid feature of the magazine as well as my second "university education." I talked with world-renowned chemists, physicists, biologists, biomedical scientists, materials scientists, and mathematicians. But the magazine folded after six years in the wake of upheavals in the publishing world, including Gabriela's departure from Springer. For a while, I continued my interviews. I talked with a total of 100 Nobel laureates and about the same number of scientists of the same caliber who were not Nobel laureates. At least not at that time; ten of them eventually did become Nobel laureates.

When did you begin interviewing scientists, and why?

In 1965, at the request of Radio Budapest, I recorded a conversation with the Soviet physicist Nikolai Semenov, a Nobel laureate, when he visited Budapest. I did not know anything about interviewing, but he was a seasoned interviewee, and it became a superb interview. It was broadcast repeatedly and published in print, and it won an award. But most of my interviews have been post-1990.

What are some concepts and practices that you have used to unify the sciences and create scientific communities?

My interest in symmetry and my editing large volumes about it gave me an excellent opportunity to bring together scientists, mathematicians, humanists, and artists in the same project. I could meet people and make friends well outside my field. I organized my first big, edited volume in the mid-1980s. I began doing this in Storrs and for the first time, I could use the freedom of sending manuscripts anywhere I pleased.

What were some of the challenges you faced in crafting careers that blend research and exposition?

It is about the importance of communicating science. My writing "career" began when I produced an extensive review about the physical meaning of interatomic distances in molecules determined by various physical techniques. This physical meaning depends on the nature of the interaction between the target and the irradiation and its relative time length as compared with the life span of the structure being investigated. Eventually, Magdi and I expanded this into a two-volume edited treatise of our principal physical technique, gas-phase electron diffraction. It involved a host of international experts — in a way, this was also community-building.

When our interest extended into the realm of symmetry, suddenly our "audience" expanded manyfold. Others had noticed that I had some ability to explain complex scientific concepts in a way that could reach interested nonspecialists. In 1986, when I was a visiting professor at the University of Texas at Austin, I was asked to teach a science course for non–science majors, and I built it around symmetry. I appreciated the increase in remuneration but did not see the significance of such a course until one of my colleagues explained it to me. He told me that it would be the most important course I would ever give, because it represented a challenge of bringing science close to future economists, military leaders, bankers, stockbrokers, politicians, legislators, and so on, that is, people on whom much of our future depends, including the support for science. There is this notion that while it is the scientists who make the discoveries, what society does with those discoveries no longer depends on them. However, especially with science becoming increasingly complex, scientists must help society to understand it. So, I consider my activities in science popularization to be my contribution to helping society become informed.

How did you and Magdi manage to do all that — research, interviews, essays, edit journals — and raise a family too?

Magdi and I have built our respective careers in a synergistic way in which our joint activities far exceeded what just adding our activities might have meant. Magdi was my student, and after her graduation, we continued working together. She soon developed her independent work, and in several aspects, exceeded me in research. In our "extracurricular" activities, we continued to share projects. Raising our two children fell much more on her than on me, although I helped in some household chores, but already the expression "helped" is revealing. However, her not becoming separated from science even when our children were small was something we accomplished together. Almost from their birth, we involved the children in our activities. When we created a book on symmetry for children, they participated — perhaps they should have been listed as coauthors. Magdi was elected a member of the Hungarian Academy of Sciences, a rare distinction for a female scientist in Hungary. Looking back on our careers, Magdi's achievements, on a relative scale, are greater than mine. She could serve as a role model to many, including me.

Thank you, Istvan, Magdi, Balazs, and Eszter! I hope that the list of your contributions to the Mathematical Intelligencer will continue to grow (see Table 1).

Table 1. Articles by István Hargittai published in *The Mathematical Intelligencer.*

Will Others Now Commit Luzin's "Sin" 45:1 (2023)

Memorials of Mathematicians in Moscow (with Magdolna Hargittai) 42:3 (2020)

The Last Boat from Lisbon: Conversations with Peter D. Lax 32:3 (2010)

John von Neumann Stamps (with Magdolna Hargittai) 28:4 (2006)

More on the ROTAS magic square (with Aldo Domenicano) 24:3 (2002)

Homage to Emmy Noether (with Magdolna Hargittai) 24:1 (2002)

John Conway–Mathematician of Symmetry and Everything Else 23:2 (2001)

Alphabetic magic square in a medieval church. (with Aldo Domenicano) 22:1 (2000)

A great communicator of mathematics and other games: a conversation with Martin Gardner 19:4 (1997)

Symmetry and perception: logos of rotational point groups induce the feeling of motion (with Magdolna Hargittai) 19:3 (1997)

Moscow Möbius. (with Lev V. Vilkov) 19:3 (1997)

Lifelong symmetry: a conversation with H. S. M. Coxeter 18:4 (1996)

Sacred star polyhedron 18:3 (1996)

One-dimensional space groups (with Magdolna Hargittai) 18:2 (1996)

Fullerene geometry under the lion's paw 17:3 (1995)

Octagons abound 17:2 (1995)

Symmetry of opposites: antisymmetry (with Magdolna Hargittai) 16:2 (1994)

Imperial cuboctahedron 15:1 (1993)

Pentagonal decoration in Granada 15:2 (1993)

Quasicrystal sculpture in Bad Ragaz 14:3 (1992)

The following images appeared in the published interview:

Left: Istvan, Balazs, Magdolna, and Eszter Hargittai in an American restaurant, 2008, by unknown photographer. Right: István and Magdolna Hargittai in Stockholm, 2008, by unknown photographer.

Left: Istvan Hargittai with John Conway (right) in Conway's office at Princeton University. Date and photographer unknown. Right: Istvan Hargittai with Martin Gardner (left) in the Gardners' home, Hendersonville, North Carolina, 1996, by Magdolna Hargittai.

Left: Istvan Hargittai with Peter Lax (left) at the Hargittais' home in Budapest, 2005, by Magdolna Hargittai. Right: Magdolna Hargittai with H. S. M. Coxeter in his office at the University of Toronto, 1995, by Istvan Hargittai.

21

INTERVIEW OF ACADEMIA EUROPAEA (AE) WITH ISTVAN HARGITTAI IN SPRING 2023 ON THE PUBLICATION OF ISTVAN HARGITTAI AND BALAZS HARGITTAI, *BRILLIANCE IN EXILE: THE DIASPORA OF HUNGARIAN SCIENTISTS FROM JOHN VON NEUMANN TO KATALIN KARIKÓ*

In this interview, conducted by the Cardiff Knowledge Hub of AE, Professor Hargittai discusses his motivation for writing his latest book *Brilliance in Exile*, which explores the lives of Hungarian émigré scientists who have made significant contributions to science after leaving their country of birth. Reproduced with permission from AE Cardiff Hub.

Professor Hargittai, congratulations on the publication of your new book. What was your motivation in writing it?

Thank you very much. Over fifty Hungarian émigré scientists are presented in this book, all having contributed significantly to science

Istvan Hargittai, 2019, lecturing at the Library of the Hungarian Academy of Sciences, photograph by and courtesy of Klára Láng.

after leaving their country of birth. I have always been interested in the nature of scientific discovery and in what makes a scientist a discoverer. I wrote a book about the Nobel Prize (*The Road to Stockholm*) for Oxford University Press. Soon after, they invited me to write a book about the five famous Hungarian scientists, the so-called Martians, who had to flee Hungary and then Europe, and went on to contribute greatly to the defense of the United States and the Free World in WWII and the Cold War. The book, titled *The Martians of Science*, was also well received. I then expanded my interest to a broader circle of émigré scientists, working with my son, Balazs. There have been several waves of scientists leaving Hungary over the past century and a quarter. The motivation behind this new book, *Brilliance in Exile*, was to show the contrast between two kinds of society. One that pushes away its gifted by intolerance, neglect, even persecution, and the other that is receptive, tolerant, and welcoming.

What have been the main reasons behind the waves of emigration by scientists, that you describe in your book? Is there some underlying reason that they share?

First let's take stock of the five waves of emigration:

1: Early 1920s: Fleeing from a virulently antisemitic regime
2: Late 1930s: Last-minute escapes from the approaching Nazi threat
3: 1945–1947: The post-war trauma of WWII and the approaching Soviet-style dictatorship
4: 1956: A brutally suppressed anti-Soviet revolution
5: 1957–1989: Escape from the confined and restrictive communist "Paradise."

The shared feature of the five-plus waves of emigration has been the desire to live in an open, democratic society where one can think and work freely. For Jewish scientists, it also often meant, literally, survival. During the past three decades, this exodus has continued and the only reason we don't call it emigration is because now travel is not restricted, so return is possible.

How did these emigrant scientists adapt to their new destinations, and ultimately succeed?

Of course, we are more aware of the success stories than of the failures. They adapted well and outperformed their prior achievements. It was not only that the open, democratic society stimulated them but also that the new environment made them feel that they had to prove themselves. The new environment in the United States and Britain tested them, but the Hungarian émigré scientists had an advantage over other immigrants in that they had already gone through a similar process when they left Hungary and started a new life in Germany (under the democratic Weimar Republic). Among the Hungarian émigré scientists, the Jewish scientists had an additional "advantage" as far as prior experience is concerned, as back in Hungary, they were often made to feel like immigrants in their own home country.

How would you characterize the relationship between the members of the scientific diaspora and their home nation of Hungary? Did they maintain a strong sense of cultural identity and belonging, despite being away?

Most émigré scientists felt Hungarian, maintained their language and interest in Hungarian literature, and when conditions became favorable, they enjoyed visiting their birth country. They related to Hungary better than Hungary related to them. Often, émigré scientists became 'non-persons' in Hungary. However, if they became hugely successful, for example, received a Nobel Prize, Hungary then treated them with pride and showered them with recognition.

What has been their position with regards to a possible return to their native country? Has there been, historically, a desire to return and introduce changes to the conditions that initially caused them to leave?

None of the great scientists returned to Hungary and no changes were introduced to the conditions that initially caused the scientists to leave. Rather, they had become rooted in their new home country, where they were treated well, with dignity, and where they were provided with close to ideal conditions for research. Also, they must have realized that returning and attempting to induce change would be an exercise in futility.

22

CONVERSATION WITH MICHAEL L. MILLER, CENTRAL EUROPEAN UNIVERSITY, BUDAPEST, SEPTEMBER 21, 2023, IN FRONT OF A LIVE AUDIENCE (EXCERPTS)

The occasion was the book launch of Istvan Hargittai, Balazs Hargittai, *Buried Glory: The Diaspora of Hungarian Scientists from John von Neumann to Katalin Karikó*. Budapest–Vienna–New York: Central European University Press, 2023. Reproduced with permission from Michael L. Miller.

During the past one hundred and twenty years, at least five different regimes existed in Hungary, and all had waves of emigration of scientists. Why is it that Hungary has produced so many great scientists and why is it that there was always a push to go

Istvan Hargittai and Michael L. Miller (right), September 21, 2023, at the Central European University, Budapest. Courtesy of Cecilia Balint, Central European University Press. Behind Miller, on the screen, Gábor Bojár, founder of Aquincum Institute of Technology, Budapest.

abroad and maybe there was also a pull? What ties together these waves of emigration and these scientists who had gone into exile?

We talk about emigration, but this does not strictly apply to the last 30 years because now return is possible and departure does not have that finality that used to be the case. This is so in principle though the movement is still mostly one-way. At the beginning of my interest in great scientists, it was not about Hungarian scientists. I wrote a well-received book about the Nobel Prize and Oxford University Press invited me to write about the Martians of science. At first, I declined but, fortunately, they did not accept it. I declined because there were others that knew the topic better than me. Besides, I thought that everything already had been written about those Martians, so there was nothing new to discuss. I tried to direct OUP to experts, but they did not need experts, they wanted another successful book after my *Road to Stockholm*. So, I embarked on a two-year study, I read everything what had been written about the Martians of science, studied archival material, and wrote the book *The Martians of Science*. Then, I did not stop there, I wrote a biography of Edward Teller who was the most controversial among the five Martians. As things were developing my interest kept expanding toward other Hungarian émigré scientists and had extensive discussions with Balazs. We came to the obvious observation that emigration was not limited to one period and one group; rather although it has happened in waves, it has been a continuous phenomenon. This is what we show in the book *Brilliance in Exile*.

The starting point was the *Martians of Science* about five Martians. George Marx who wrote extensively about the Martians defined them as Hungarians — and not just scientists — who went to the West and became famous. We thought this to be a euphemistic usage of the term, masking the fact that these were exiled Jewish scientists. We show why it is justified to call it exile rather than a choice of adventure to see the world as Marx often implied. We have adhered to the definition of the Martians who were great scientists who were willing to risk their scientific careers during the Second World War and the Cold War to defend the United States and the Free World. All five came from upper-middle-class Jewish families in Budapest.

The continuous stream of emigration was a consequence of the actions of the political systems, one after the other, that wanted to get rid of those who were involved with the previous regime or were suspected that they were. This was present in the Horthy regime with respect to the 1919 communist dictatorship and in the socialist system with respect to the Horthy regime. Then, after 1956, those who participated in the anti-Soviet revolution or were suspected of participation were blacklisted and were not allowed to teach, to educate young people. A foreign visitor once remarked to the secretary of the Hungarian Academy of Sciences that Hungary must be a very rich country to afford wasting so much talent.

Each political system during the twentieth century had its favored people and its enemies or perceived enemies. Compare it with the British system where, ever since the seventeenth century — from the time of the Restoration when there did exist some political discrimination among scientists — there has not been any consideration of political background in professorial appointments or in the elections to the Royal Society. Advancement in academia and politics should not mix and they never did in Britain, and they always have in Hungary. There is then the question of whence so many gifted scientists? I don't know but it appears that however many leave, plenty of new ones follow and this continues though we can't be sure that this will keep happening indefinitely.

I did some research on the Bauhaus movement in Dessau, Germany, and elsewhere. There were a dozen or so Hungarians and there was a question whether there was something especially Hungarian that made their presence so conspicuous in the Bauhaus movement, the designers, for example. I'm asking the same question about the scientists. Is there something that is distinctively Hungarian in science and the results that are produced, the methods that are used, something that could be tied back to something Hungarian? You have 50 individual chapters but you deal with more like a hundred scientists.

It's wonderful when you mention architects and designers. In school, we learn about how to handle perspectives in drawing. Once I talked with a visiting designer from New Zealand who came to study how perspectives are being taught in Hungarian schools, someone had suggested that perspectives are taught differently in Hungary from the rest of the world. She was trying to correlate this way of teaching and the relatively large number of Hungarian designers and their unique approach in their work. I have no idea whether there is anything solid in this notion. What I can say is that even on this anecdotal level there is nothing in our science education that would point in any direction of peculiarity conducing to later achievements. Except, perhaps, that the Hungarian high school, the *gimnázium*, used to be excellent, truly outstanding. It was, that is, around the turn of the nineteenth and twentieth centuries. It is still quite good although it has been deteriorating for decades. How good it could have been a hundred and twenty years ago. What I can suggest as one of the contributing factors is this: When these Hungarian scientists, mostly Jewish, found themselves in a new environment, first in Germany, then, in the United States, we can compare them with other émigré scientists there, from Germany, Belgium, Italy, and other countries. The Hungarians had the "advantage" that they had already been seasoned émigrés in going first to Germany. The Hungarian Jews then had yet a further "advantage" in that they were made to feel themselves strangers in their homeland. This expression I borrowed from the émigré historian Istvan Deak. When they arrived in the United States, they accommodated themselves more easily than the others and could focus from the first moment on doing creative work.

Part Two

Selection of Forewords and Reviews

I

ISTVAN HARGITTAI, MAGDOLNA HARGITTAI, *SYMMETRY THROUGH THE EYES OF A CHEMIST*, 3 EDITIONS, 1986 (VCH), 1995 (PLENUM), 2009 (SPRINGER).

A quote from the review by David R. Rosseinsky, University of Exeter, "Fearful symmetry." *Nature* April 30, 1987, 326, 908.

"Almost education by stealth, certainly education by aesthetic appeal."

***

Review by Philip Morrison, Professor of Physics, MIT, "The origin of shape." *Scientific American* April 1987, 256, No. 4, 22–23. This is a joint review with the volume of Istvan Hargittai, ed., *Symmetry: Unifying Human Understanding*. Oxford: Pergamon Press, 1986. Reproduced with permission. Copyright © 1987 Scientific American, Inc. All rights reserved.

A witty cartoonist shows us the atelier of the Creator, who is judging with pleasure the wall where young angelic designers have posted their competing projects for the decorated hexagons that will become snow crystals: "Impossible to choose; produce them all."

The cosmopolitan eye caught by the Parisian cartoon, good-humored and clearly delighted by diversity, informs this entire book. The work offers a broad new perspective that looks outward over the rich and often demanding monographic literature. Its authors are two structural chemists, research workers at the Hungarian Academy of Sciences in Budapest, who have seen everything relevant. They show most of it to us in photographs and drawings and tell us about it lightly and well. As an instance, their figures present clear although spare prototypes of the 17 plane groups that describe all possible repeating wallpaper patterns; each array of black triangles is accompanied by a strictly corresponding example documented from Hungarian needlework. All 17 groups are found in that needlework, as well as all seven classes of the border decorations that are their one-dimensional analog.

The book has three roughly equal parts. The first part sets out one after another the symmetries whose combinations impose such interesting limits on form. The opening sections are studded with examples, both fresh and familiar, from the Van Dyck portrait of Charles I painted in full face and in both profiles, through a photograph taken by her chemist-parents of little Eszter Hargittai along with her reflection, to the intricate Pueblo pottery that has only rotational symmetry.

Inversion, polarity, and chirality (handedness, as in a DNA molecule) are treated at some length, now with molecules in mind. Two closely related organic crystals are shown, the pointy one anchored by a center of symmetry, the other smugly lacking it: a gas-solid reaction corrodes one face of a narrow slot cut in a crystal distinctly faster than it eats at the other. Drawings and tabulated properties of the five regular convex and the 13 semiregular polyhedrons are also presented; they serve both as part of the classical lore and as examples for discussion of symmetries and their notation. Kepler's fit to the planetary orbits is examined along with the goodness of the fit. The lists of references are long and useful. They offer the reader the simple canons of the subject, for example, Hermann Weyl's masterly popular lectures, Martin Gardner's writings and Edwin Abbott's Flatland — as well as up-to-date textbooks and technical papers.

The shapes of molecules are the next broad topic. A disarming and unforgettable start is made in the account of the standard projectional representation of isomers that differ only in the relative orientation of their parts. Two drawings after Degas are invoked: one dancer with arms flung apart at the end of an arabesque, who shows the staggered conformation; the eclipsed conformation is embodied in another dancer, who bends to adjust her shoes.

Cubanes and pentaprismane have been prepared, hydrocarbon molecules shaped as they are named. Adamantane is a hydrocarbon analogous to diamond, whose carbon framework resembles a set of four cubes, one inside the other. It is a real substance; even multiply joined adamantanes have been synthesized. (The book includes an index of formulas that is a page and a half long. Most of the entries are organic molecules, but there are plenty of boron and fluorine compounds as well.)

This pragmatic and qualitative scheme is rather faithful to the results of genuine ab initio quantum calculations for a wide class of molecules, those dominated by a few central atoms. The first part of the book ends by treating electron-cloud shapes in molecules according to a simple and popular method known somewhat Germanically as the valence-shell electron pair repulsion model. An alternate name for the model suggests one reason: a recent proposal sought to recognize explicitly the role of the exclusion principle by calling the scheme Pauli mechanics.

The closing third of the book extends the ideas of combined symmetry from operations about one point to those that build infinite lines, planes, or volumes. We have already mentioned the border and plane symmetry groups. Next, there is a useful introduction to classical crystallography, sweetened by stacks of cannonballs and the old cleavage figures of the pioneers. There is also a dazzling long quotation (illustrated with his own drawing) from the writer Karel Čapek describing a visit to the minerals at the British Museum. "Even in human life there is a hidden force towards crystallization. Egypt crystallizes in pyramids and obelisks, Greece in columns; the Middle Ages in vials; London in grinny cubes ..."

Easily understandable drawings of crystal forms and the corresponding projection diagrams illustrate the 32 crystal point groups. The restrictions imposed by combining rotations, mirror reflections, and translations are nicely described. It is easy to grasp why by these rules nothing mineral is fivefold. The text is sufficiently contemporary to present a brief comment on the recent discovery of a quasicrystal or two with a penchant for fivefold symmetry. Our authors are far from dismayed, for the tendency of their final chapter is to view classical crystallography, with its rules derived from the mathematics of infinity, as an altogether too restrictive framework. The moderns could be counted on to depart from that hog-tied posture of perfection as materials of new kinds were made and examined.

The helical molecules of life are discussed (but rather too briefly), and there is something about the viruses as finite versions of symmetrical packed structures. The full 230 crystal space groups are presented; wisely they are not tabulated in detail.

The last sections begin to add notions of energy to the deep accounts of form. The Soviet physicist A. I. Kitaigorodskii has for decades treated molecular organic crystals as examples of packing. Using sphere close packing as an introduction, he examined the ways molecules of arbitrary shape could be packed into lattices under the various space groups. It is not always possible to fit molecules close together without overlap and without allowing some convex surfaces to face other convex surfaces. Clearly such packing schemes cannot have high molecular density. Molecules with some added symmetry of their own (for example a plane of symmetry) admit closer neighbors even in unfavorable space groups. From such arguments, Kitaigorodskii shows that in only six of the 230 space groups could planes of molecules be formed that have six neighbors in contact. Just those few groups describe the bulk of the known organic crystals. The classical groups are geometrically equal; energetically they differ, even grossly.

In the middle third of the book the treatment suddenly becomes abstract. Most illustrations are diagrams and graphs; virtually gone are the scenes from the real world. It is a helpful introduction to the algebra of group representations, complete with matrix multiplication, reducibility, and group character tables. This expertise is not carried to the point of much calculation, yet most readers will hardly make their way smoothly past a projection operator without having paper and pencil at hand.

It is the dynamics of molecules that flow from this excursion away from visualization into algebra. A fine detailed look at the normal vibrations of the water molecule provides a halfway house. The text then pursues up-to-date ideas about the use of symmetry to understand single-step reactions between molecules. The scheme is to take some account of the tendency for merger in collisions that match the symmetries of the approximate wave functions. These constraints lead to general rules in simple collisions. Symmetry-related rules of reaction have been growing in utility for two decades, and the Nobel lists recite the names of their formulators: Fukui, Hoffmann, and Woodward. This part of the book is too technical to win many general readers.

The Hargittais wrote in English, mostly during a two-year stay at the University of Connecticut at Storrs. The book had a modest progenitor in Hungarian published by the academy as part of a paperback series in popular science. It has since had a grander and bulkier issue as well: the ambitious symposium on symmetry, orchestrated by Istvan Hargittai, to which more than 70 mathematicians, artists and a variety of scientists from many lands contributed.

I cannot profess to have attended in full to the 1,000-page volume with so many articles of differing levels, topics, and length, from particle physics to country dance; yet a very few pieces quickly caught this rather daunted eye. Musical and literary symmetries are here. Two papers generalize the Kekulé benzene ring, a carbon-atom hexagon in which alternate links are doubled. The puzzle of how many ways there are to join benzene rings has already led to a voluminous literature and some remarkable formulas. The phenomenon of moiré is described with interesting works of moiré art; the simpler examples plainly invite more mathematical analysis.

The big regular jointed basalt columns found in Fingal's Cave and in similar showy formations here and there are hard to explain. Their strange pseudocrystallinity is often accounted for by a theory of the 1880s, which proposes a process of growth so unstable that it would require perfection. Now there is a new understanding, if not yet a full answer. At their base the columns are four-sided, displaying the same kind of irregular T joints as those seen in surface-tension cracks in mud and in some pottery glazes, not the Y joints of the hexagonal mesh that become so conspicuous well above the base. Here the ideal stress-relief pattern is being approached through the third dimension.

2

ISTVAN HARGITTAI, MAGDOLNA HARGITTAI, *SYMMETRY: A UNIFYING CONCEPT* (BOLINAS, CALIFORNIA: SHELTER, 1994)

Review by Philip Morrison, Professor of Physics, MIT. *Scientific American*, December 1994, 123. Reproduced with permission. Copyright © 1987 Scientific American, Inc. All rights reserved.

A globe-trotting couple, structural chemists from Budapest, have assembled this fine display for general readers from ages 12 to 120. It shows the world of symmetries the authors first entered through the austere portals of electron diffraction.

First of all, it is a book of pictures in black and white. Half the book considers and displays in plenty of dance masks, chiral pairs of gloves and artfully dissected apples ("The Royal Cut", the French call it), pinwheels and flowers, six-fold snowflakes and the Eiffel Tower (a four-fold blossom as seen from straight above), balloon clusters, even molecules. The latter half goes beyond mirrors and rotations, all those moves that leave one point fixed, to look at grander repeats, in fences and friezes, in spiral and helix, honeycombs and tiled walls, and finally in diamonds, Eschers, and even unsettlingly novel alloy quasi-crystals so full of neatly arranged pentagons. (Of course, they also show the Pentagon itself.)

Art and nature draw similar attention here, mostly at the scale that meets the eye. The authors' own special interest in the micro-symmetries inside matter is here background, more strongly seen in their other half-dozen more technical books on the topic. As a field guide to easily grasped and beautiful exemplars of this powerful idea, the book is a storehouse hard to beat.

A few pages in the middle between point groups and space groups touch on the anti-symmetries of this world, not the merely fuzzy symmetry of left hand and right, but black for white, negative and positive, concave and convex, even angels and devils. Here's room for a whole new book!

3

ISTVAN HARGITTAI, MAGDOLNA HARGITTAI, *IN OUR OWN IMAGE: PERSONAL SYMMETRY IN DISCOVERY,* NEW YORK: KLUWER ACADEMIC/ PLENUM 2000

Foreword by Harry Kroto[1]

Symmetry is an endlessly fascinating area of aesthetic intellectual activity. Appreciation of symmetry seems to go back to the earliest time of human intellectual development as can be seen from the artefacts that have survived from our early ancestors. It may even go back much further — deeper into the way our nervous system responds to stimuli — optical, oral, etc. After all, many flowers have high symmetries and, hence, the sensory systems of insects and birds must be tuned so that they are stimulated by symmetric structural and associated color patterns. Perhaps some observed patterns map directly onto neural sensors and in some way relate to the physical structures of the associated networks. Symmetry may actually be an essential all-pervasive fundamental aspect of perception by any sensory mechanism.

---

[1] Harold (Harry) W. Kroto (1939–2016), Nobel laureate in Chemistry.

This book approaches symmetry in a novel way — it is an entirely personal perspective of the authors, Istvan and Magdi Hargittai, on many individual personalities that punctuate key moments in the history of science — in particular, the scientific advances with these obvious symmetry-related aspects. The Hargittais survey the history and project out numerous fascinating insights from a symmetry-related vantage point and show how patterns may influence the ways in which scientific and technical advances are made.

This book takes us through some of the most fascinating areas of science with the aid of numerous accounts by outstanding scientists as well as related anecdotal information. Difficult scientific advances are often discussed from a general conceptual viewpoint in a way that makes them less forbidding. This will help readers to deal with these advances if they wish to understand them in detail at some later stage. Often, we are treated to discussions obtained firsthand from the scientists who made the groundbreaking advances or from others who were involved indirectly. These accounts invariably reveal fascinating new information.

The black and white illustrations by Orosz are a welcome change from the present over-reliance on photographs and/or computer graphics which have their place — but not in this volume. Orosz has brought a measure of "personal" imagination and creativity which complements the text well and enhances the book's general intellectual appeal immeasurably.

I thoroughly enjoyed the spirit of the book and learning more about numerous colorful scientists. The use of symmetry as the coat hanger on which to hang the stories and correlate the plethora of personal observations has enabled the Hargittais to create a book which is not only unique but also an invaluable addition to our literary heritage.

***

A quote from the review by David Blow, Professor of Biophysics, Imperial College London, "The bondage of symmetry broken." *Nature* August 24, 2000, 406, 828–829.

"Hermann Weyl's classic book *Symmetry* concentrated on the interplay between the mathematical operation of symmetry, the success of the artist, and the satisfaction of the beholder. In delightful contrast, Istvan and Magdolna Hargittai emphasize the deadness of perfect symmetry, citing Pierre Curie's 'dissymmetry makes the phenomenon,' and Fedorov's 'crystallization is death.' They follow, in fascinating detail, lines of development where the relaxation of an imposed or imagined symmetry has led to the deepening of scientific insight."

4

ISTVAN HARGITTAI, *CANDID SCIENCE: CONVERSATIONS WITH FAMOUS CHEMISTS*. EDITED BY MAGDOLNA HARGITTAI, LONDON: IMPERIAL COLLEGE PRESS, 2000

Foreword by George Porter[1]

This collection of interviews by Professor Hargittai is a timely celebration of the remarkable century of chemistry, which has just passed. Those interviewed are some forty chemists, who were fortunate enough to live during the momentous developments in chemistry, which have occurred during the last decades of the millennium.

These historic events will be retold for centuries to come, often with more understanding and appreciation than we can have today. But, in one respect, what is written here will never be repeated because it is the personal experiences of the chemists who took part, each of them a unique individual.

Scientists are sometimes seen as automata without human feeling or sentiment (unlike artists for example!). The scientists themselves, and those who live with scientists, know how false this view is. But

---

[1] George Porter (1920–2002), Nobel laureate in Chemistry.

unfortunately, most people know very little of science or scientists personally and often see them as living in a different world from "ordinary" people.

I hope that the interviews recorded here will do something to dispel this increasing divide in our understanding. They tell how chemists of today have lived and worked, why they chose chemistry as a career, and how they were led to make the discoveries for which they are best known. The informality of the interviews gives the impression that one is sharing in a conversation rather than listening to an autobiographical lecture.

Children are born scientists but, if they are to retain their love of science, they need role models to follow and encouragement from those who have been successful. This book will be enjoyed by all who have some interest in science, and it will be of special value to the young people whom it may encourage to follow those, whose stories are told here.

***

A quote from the review by Gautam R. Desiraju, School of Chemistry, University of Hyderabad, "Reactions of a chemical kindred." *Nature*, June 29, 2000, 405, 996–997.

> "Research is and will always be exciting, but the conversations in this book encapsulate a time that is past and leave the reader with a comforting glow. The main protagonists have told their tales, and the editor has conducted his interviews with sympathy and collected his material with care. For this, he is to be commended. His book will be enjoyed by chemists and non-chemists alike."

***

A quote from the review by Dennis Rouvray, Department of Chemistry, University of Georgia, Athens, "Talking chemistry." *Chemistry in Britain*, September 2000, 56.

"One comes away from this book with the impression that one has been listening in on a conversation between two old friends: the tone is relaxed, the setting is informal, and the information exchanged is fascinating. So fascinating in fact that I found this book exceedingly hard to put down. Because such intimate revelations are usually not made public, this work constitutes a valuable historical document and is one that I feel should be read by every chemist."

***

Review by David Knight, Professor of History and Philosophy of Science, University of Durham, UK, "Candid Science." *Chemistry & Industry*, March 5, 2001, 149–150. Reproduced with permission from Neil Eisberg, Editor of *Chemistry & Industry*.

The cover of this wonderfully personal and accessible book is decorated with 36 photographs of eminent chemists. From academe to industry, Istvan Hargittai has assembled their candid comments here. Clearly, he is brilliant at drawing his subjects out, tactfully asking them the right questions and thus providing us with an insider view of what chemistry in the 20th century has been like.

We come up against peer review, patronage, and personality: chemists have never been desiccated calculating machines, so that (along with the admirable drive to understand) ambition, quarrels, priority disputes, and competition for honors, resources, and prestige go back to Lavoisier's time. Self-analyses, reminiscences, and unguarded comments on the style, successes, and failings of eminent chemists make this compulsive reading; we rarely encounter the bland style of the formal obituary.

If there is a pattern, it might be that chemists who have become eminent often seem to have rather broad interests. They have responded to adventitious circumstances, taking advantage of the facilities available where there happened to be a job going. Being in the right place at the right time, or making the place and the time right, is vital.

Some of those interviewed died before the book (based on articles in The Chemical Intelligencer) was published, which indicates the importance of getting on with such oral history.

While some interviews are predominantly personal, others have a more chemical or institutional focus. If we thought that discovery was always a simple process, a eureka moment that befell an individual, we would be put right here: there are several competing accounts of what took place as experimental and theoretical insights gelled. We get these multiple perspectives, particularly over the story of the fullerenes, where a hundred pages of the book are devoted to the various participants' stories; and we see that there is no simple answer to the question 'what happened?' Here as elsewhere the history is complex, but each telling casts more light upon it; no doubt different readers may come to different conclusions about where credit is due — or which are the most important factors — and duly sympathize with the committees choosing where to award Nobel Prizes.

Whereas 19th-century chemistry was dominated by Germany, in the 20th century the US became the chemical superpower. Some great scientists, such as Pauling (with whom the book begins) were born and bred in the US; but many of those interviewed made their way there, either as refugees or as economic migrants, going down the 'brain drain' from Britain and other countries. Quite a high proportion is Jewish.

We see evidence for the great change that has come over science in the past 50 years or so with instrumentation. Methods of teaching and supervision (on which there are interesting comments), as well as of research, have been transformed by the new scale of things. Practical skills like glassblowing and even weighing, character-building exercises involving Kipps's apparatus and qualitative-analysis tables, and know-how about smells and colours, have given way to measurements with wonderful and expensive machines.

The money required had to be raised, and this is part of the story of US pre-eminence; we read a good deal here about research grants and how they were obtained — notably from the government.

World War II and then the Cold War generated this big science, and there are accounts of how it happened and how it works.

Women like Berzelius's famous housekeeper Anna, and the wives of other chemists, played a part behind the scenes of the 19th century. By the 20th century, they were becoming visible, and we duly meet a few distinguished women here; when a similar volume is done in 50 years, we may expect that women will be much more prominent. The women chemists in this book worked on natural products in Turkey, worked for a pharmaceutical company, and also worked in the US and other universities.

But in the generation portrayed here, women (sometimes making an appearance only as the wives of those interviewed) were expected as a rule to abandon science when they got married and move, like Anna, into the background. Their stories involve overcoming prejudices; science generally, and perhaps chemistry particularly, with its heroic experimental tradition of stinks and bangs, has a very macho image. Nature in its rhetoric is generally interrogated, mastered, conquered, and sometimes penetrated; perhaps women will introduce a more peaceful vocabulary of conversation and comprehension.

Chemistry is the science of making things out of others and has always had strong connections with industry — where the satisfactions are somewhat different from those of academe. In the 19th century, chemists got good marks for their explosives, fertilizers, and dyes. In the 20th century, following World War I ('the chemists' war') the triumphant science and its applications, while ubiquitous, have seemed more alarming. 'Chemical' has become an off-putting adjective to many, unfortunately associated with warfare, pollution, or profiteering. We don't meet that here. Instead, we read of industrial laboratories where researchers were effectively allowed to pursue blue-skies research; and interesting comments about how really successful directors managed their talented associates.

Pauling emerges as unusual in both the extent and direction of his political interests; his campaign against nuclear weapons and for social justice did not arouse much resonance within the US chemical community. Most of the chemists here seem to be non-political, probably meaning as usual mildly right of centre, and some are strongly conservative. Probably most scientists have always favored governments that fund them and let them get on with their job, and do not interfere unfavorably in the market.

The chemists interviewed mention various factors that drew them into chemistry in the first place, and Paul de Kruif's *Microbe Hunters* of 1927 is mentioned so often as to warrant a couple of pages about it.

It might now be difficult to develop idealistic enthusiasm via popular science. We are older and wiser about science and its effects, and perhaps too pessimistic about the capacity of technology to go on solving our problems. We don't find any such despondency here; these people after all are enthusiasts for it. They make the book delightful to browse. Maybe it will, as it should, produce enthusiasts in its turn; it should be on everyone's bedside table.

***

A quote from the review by Stephen J. Angyal, School of Chemistry, The University of New South Wales, Sydney. *Australian Journal of Chemistry* 2003, 56, 633–634.

> "This book, of course, can be opened on any page and read with pleasure. The personal speaking style of the text, in contrast to laboriously written paragraphs, make it easy reading; one feels that one is taking part in a conversation rather than listening to a lecture. There is plenty of useful information, interesting adventure, and valuable advice in the text. The book will be enjoyed by all who have some interest in chemistry, and it is highly recommended."

5

ISTVAN HARGITTAI, *CANDID SCIENCE II: CONVERSATIONS WITH FAMOUS BIOMEDICAL SCIENTISTS.* EDITED BY MAGDOLNA HARGITTAI, LONDON: IMPERIAL COLLEGE PRESS, 2002

Foreword by Arthur Kornberg[1]

More than ever before, the true stories of scientific discoveries are lost by the restricted space and style of the widely read journals and the hurried pace of reporting. There is no means of knowing either the personalities or the artistic qualities in the pursuit of understanding nature. Scientists look to the future rather than the past; they abhor writing and do it poorly. Autobiographical accounts are rare and the memories of the deceased in the volumes of the National Academy of Science (U.S.) and the Royal Society are commonly dull and impersonal. There is an important need to know about the people responsible for the progress of science as much as about those in politics, business, the military, and the arts.

---

[1]Arthur Kornberg (1918–2007), Nobel laureate in Physiology or Medicine.

Oral histories can supply these much-needed accounts of how discoveries were made as told by those who made them. The success of this kind of historiography is attested by the enthusiastic reviews of István Hargittai's first volume, *Candid Science: Conversations with Famous Chemists*. This second volume, devoted to thirty-six biologically oriented chemists and chemically oriented biologists, is just as good. There is fluency and intimacy, and hence readability, in the responses to questions from a knowledgeable colleague who had done his homework on both the scientist and the subject.

Hargittai poses questions that might not have been considered in an autobiography and that might have been frustratingly impossible for the author of an obituary memoir. Fascinating revelations emerge in the responses to questions that range from descriptions of the research subject, family background, inclinations to do science, religious beliefs and appraisals of other scientists, dead or alive. Even though I know many of the scientists, some rather well, I learned things about each that were instructive and gratifying.

Among oral histories, there are book-length volumes, such as the current series on biotechnology conducted by the Bancroft Library of the University of California, designed exclusively for archival use. By contrast, this volume is a selection of hors d'oeuvres to be enjoyed by a wide readership. There are of course limitations to such brief oral histories. Leo Szilard once referred to "my version of the facts"; other versions would be helpful for clarifying the record.

All told, these unique volumes, and more that have been promised, are exceedingly worthwhile and can be enjoyed by all, young and old.

***

Review by Sally Smith Hughes, Center for Science, Technology, Medicine & Society, University of California, Berkeley. *ISIS*, 2003, 94/1, 191–192. Reproduced with permission from Sally Smith Hughes.

Scientific publications and journalistic accounts seldom provide the circumstances of the reported research or characterizations of the scientists involved and their motivations. A current depersonalization of science, combined with the stereotypical depiction of scientists as social misfits, leads the public to negative views of scientists as human beings. This collection of short interviews with prominent biomedical scientists may help to dispel this image. As Arthur Kornberg describes in his brief foreword, Istvan Hargittai, a structural chemist at the Budapest University of Technology and Economics and Research Professor at Eötvös University in Budapest, interviewed thirty-six biologically oriented chemists and chemically oriented biologists. Hargittai's wife Magdolna, also a scientist, was responsible for two of the interviews and served as editor.

The interviewees are Nobel laureates or figures otherwise widely known in European and North American biomedical research. The list includes physicists, chemists, biologists, physicians, and others, only two of them women. One gathers that they were selected for interviews because of their standing at the pinnacle of biomedical science, but the selection process is not described. The interviews are illustrated with photographs from the interviewee's personal collection or taken by the Hargittais at the time of the interview. Most of the interviews were first published in *The Chemical Intelligencer*, of which Istvan Hargittai is editor-in-chief.

In some cases, one can discern from the author's references that Hargittai prepared for an interview by reading a book or books by or about the interviewee that he used as a basis for his questions. He does not give evidence of having systematically read through an interviewee's published and unpublished papers. As a result, chronologically development of a scientist's career and institutional and disciplinary context is often lacking. Although Hargittai is skilled at drawing out his subjects and sometimes elicits surprising and revealing answers, the somewhat scattershot approach to interviewing leaves the reader with a disjointed picture of individual scientific achievements and contributions to discipline building.

Although Hargittai states in his preface that he sees his role "as a fellow scientist rather than that of an investigative reporter" and admits to dropping interviewee-declined questions from the published material, he is not afraid to ask hard questions on occasion. For instance, of Benno Müller-Hill, known for work on the *lac* repressor, he asked: "According to some, despite your diverse contributions to science, it's likely that you will be remembered for one thing only. Would you mind being remembered more for your anti-Nazi studies than for your genetics?" (p. 127). (For the record, Müller-Hill responded that he hoped his isolation of the *lac* repressor would also count.) In a few instances — that of Marshall Nirenberg is one — presumably because the interview was problematic, Hargittai provides an essay rather than the interview itself.

This book is part of an interview series entitled *Candid Science* and is preceded by a first volume of interviews titled *Candid Science: Conversations with Famous Chemists* (London: Imperial College Press, 2000); it will be followed by a third volume with chemists and a fourth with physicists.

So, what does the reader gain from this collection? The answers are entertainment and further evidence that scientists have engaging stories to tell that enhance the published scientific record and reveal themselves to be fascinating beings with colorful personalities and motivations as familiar as those shared by the rest of the human species. These are important points to make in a world increasingly driven by impersonal science and technology. For students, this book provides useful overviews of the work of scientists who have helped to make biomedicine the dominant science of the late twentieth and twenty-first centuries. The historian may use these interviews as a starting point but will need to fill them out with further research before reaching any credible conclusion about individual scientific contribution and relationship to the development of his and her research field.

6

ISTVAN HARGITTAI, *CANDID SCIENCE III: MORE CONVERSATIONS WITH FAMOUS CHEMISTS.* EDITED BY MAGDOLNA HARGITTAI, LONDON: IMPERIAL COLLEGE PRESS, 2003

Foreword by Herbert A. Hauptman[1]

It has been a good many years since I read my first book on the history of science. Although I no longer remember the author or the title of that book, I do remember the impression it made on me. First, it awakened in me a deep and abiding interest in science, an interest which has endured and become even more intense with the passage of time. Then too, it has inspired me to read other books on the history of science and these in turn, as in a chain reaction, impelled me to read far more of science history than I could possibly absorb. Admittedly then, much of what I read in those early years was beyond my comprehension; however, over time, understanding eventually did follow.

There was one facet of this early activity which I found particularly fascinating and rewarding. This was the aspect of science history

---

[1] Herbert A. Hauptman (1917–2011), Nobel laureate in Chemistry.

which was concerned with the lives of the great scientists themselves, the challenges they met, the struggles they faced, and the obstacles they overcame — when they were successful. Not only are these stories of great intrinsic interest, but I believe there is no better way to stimulate the creative impulse than to learn at firsthand from the words of the masters, those whose work has survived the test of time.

Who, for example, can fail to be inspired by reading, in his own words, Fermat's statement of his famous "Last Theorem" and his claim to have found a truly marvelous proof of this magnificent theorem? An additional bonus is the insight one gains into the process which guided Fermat to his remarkable and surprising conclusion. The fact that the problem itself eluded all attempts at solution for some 350 years, finally yielding in the most unexpected way, adds an appropriate air of mystery to this extraordinary tale. The proof itself must be regarded as a major mathematical triumph, if not the supreme mathematical gem, of the twentieth century.

I remember, too, with what pleasure I read, for example, Archimedes' account of his discovery of the volume and surface area of the sphere and of Gauss' proof, in his own words, on the constructability, with straight edge and compass, of the regular polygon of 17 sides, totally unanticipated at the time. In both these I felt as if I were sharing in the joy of their discoveries, almost as if I were participating in the discovery process itself, a feeling which I think can be conveyed in no other way. I believe in fact that no third person account can possibly generate the same sensation of pleasure one derives by sharing with the author himself his own feelings of excitement as he recounts the circumstances of his discovery.

These examples then provide, in my opinion, convincing evidence that an important purpose is served by those who take upon themselves the task of scientific biography which attempts to throw light on the origins of discovery, as Professor Hargittai attempts to do in his book. I believe, too, that much of the earlier biographies which already exist were written after the death of the scientist involved, often long after, so that no first-hand account is available. Then, too, these earlier biographies were often written by those with little or inadequate scientific background so that the circumstances and the full significance of the scientist's contribution were not always made clear.

Professor Hargittai's contribution, on the other hand, serves the unique and important purpose that it presents in a timely way, while the scientist himself is still alive, and in the first person, the significance of his contribution and the context in which it was made, as he himself has seen it. For this reason, the reader himself feels a closer connection with the scientist than otherwise would be possible and develops a deeper appreciation of the author's contribution than he otherwise might. In any event, this has been my own experience, and I believe that Professor Hargittai's book will serve the same function, the importance of which cannot be exaggerated, since I believe it will stimulate the reader to think in new directions.

***

A quote from the review by Arthur Greenberg, University of New Hampshire, Durham, "Chemical Conversations." *Nature*, June 12, 2003, 423, 687–688.

> "*Candid Science III* provides snapshots of the culture and progress of chemistry over the final 60 years of the twentieth century."

***

Review by David Knight. Professor of History and Philosophy of Science, University of Durham, UK. *Chemistry & Industry*, June 16, 2003, 22–23. Reproduced with permission from Neil Eisberg, Editor of *Chemistry & Industry*.

How do we get into other people's heads, find out their feelings about what they have done, and how they see their life, family, friends, and colleagues? Well, we talk to them: aware that nobody is on oath in their conversation, nor wholly candid, and may be tidying up and sanitizing the past, making their role more or less heroic than it was. István Hargittai's interviews with eminent scientists are a lively way of entering the minds of prominent chemists. Previously, often we have had to wait until chemists are dead to find out what they did, and in obituaries, we have an outsider's view, equally valid in its way but not the same.

In this volume, there are 36 interviews, each accompanied by a portrait (reproduced in miniature and color on the cover), with chemists who have mostly retired — or reached retiring age, because several are clearly unstoppable. They have had the chance to reflect on their lives and careers as scientists; sometimes to be catty about colleagues; and, to think about the chances and changes in life. There is a joke about the way to make God laugh is to tell him your plans; that is exemplified here, though there are a few whose scientific life does not seem to have gone according to plan.

Some of these conversations have been reconstructed as a block of prose, occasionally in the third person, which is a bit dull. But the loveliest are set out as dialogues, with good questions (often personal and occasionally verging on the impertinent) and answers that illuminate the way 20th-century chemistry worked. Clearly, much of it happened in North America.

About a third of Hargittai's conversationalists were immigrants, involuntary refugees, or going down the brain drain — in the tradition of Joseph Priestley. But whereas 200 years ago he was leaving Europe, the center of science, for the banks of the Susquehanna River, these were going to new centers of excellence, to find equipment, funding, and stimulus in the new world. Many were Jewish: few of them and those brought up Christian, seem to have retained a religious commitment as Priestley did, and some comment on how it is to be agnostic in the U.S. In Europe (where some interviewees live and work) there is no problem presumably.

Scientists emerge as quirky, some coming across as attractive people and some not, some driven and others more relaxed, but all bubbling with enthusiasm for chemistry and, on the whole, happy that their work has been interesting and fruitful. There is nobody such as Lord Kelvin, the great classical physicist, reflecting a century ago at his jubilee party that one word characterized his work: failure! He was pondering the paradoxical luminiferous ether, and no doubt he was tongue in check. But optimism is a feature of this book, and the participants seem confident that they have the right end of the stick.

As well as learning about 36 eminent chemists, many of them Nobel laureates and others who might well have been, we read about their interactions, sometimes with each other. Reading two sides of a story is always fascinating, and uneasy relationships feature often in the accounts. This should not surprise us, for scientists' ideas and particular techniques are, after all, their most important asset; science is a competitive as well as collaborative activity.

John Polanyi (p 384) has interesting thoughts about difficult audiences: "When you go to a scientific meeting with a new idea, you do not expect people to applaud. What they do is to tear it apart. That is their function. In science, we arrive at truth through adversarial dialogue." People forget this when, for example, they look back at events such as the Huxley–Wilberforce debate on evolution. Converting people (editors of journals and their referees, directors of laboratories and members of funding committees, as well as colleagues and competitors in the same field) to new ways of looking and doing is an important theme in this book. Being in the mainstream is vital; occasionally, given a bit of luck, diverting it your way and watering a whole new field.

We meet a number of women, who have interesting comments about their opportunities and problems, and chemists whose career has been industrial rather than academic — though many academics have industrial connections. Some of the participants discuss institutions, such as the French Academy of Sciences, for example, from an insider's perspective. Others reflect on the award of Nobel Prizes, and why some eminent chemists never received one. We meet some chemical popes, in the tradition of the inerrant William Wollaston of 200 years ago, and some dogmas, noted where, instead of usefully focusing attention on profitable questions, they blinkered scientists who should have opened their eyes wider.

We learn also about how chemistry, this vast body of knowledge, has developed in the past half-century, the active life of these people. There is a good deal of discussion of the way computation threatens or promises to take over from experiment, and of instruments that have superseded the test tube. Thus, on page 383, there is a picture of a chemistry laboratory in Toronto that looks similar to an engineering workshop, although on page 226 Árpád Furka reassuringly does chemistry with glassware.

We note the prominence of medical connections; the tremendous advances in determining molecular structures, and the importance of time in elucidating reaction mechanisms — for reactions may be very slow or fast. Altogether, this makes a wonderful bedside book that attractively opens up a gallery of people and a window onto the development of science.

***

Review by Steve Ritter, "Sitting Down with Famous Chemists." *Chemical & Engineering News*, May 12, 2003, p. 38. Reproduced with permission from Mitch Jacoby, Executive Editor of *Chemical & Engineering News*.

Like in any profession, scientists can learn an incredible amount, and in turn be introspective, by talking with other scientists. But when you don't have time or a leading authority on a field of study down the hall, what do you do? One answer is to turn to Candid Science III: More Conversations with Famous Chemists, the latest in the series of "conversations" books put together by Hungarian chemistry professor Istvan Hargittai of Budapest University of Technology and Economics.

Hargittai uses his interviews to take us inside the personal and professional lives of famous chemists as though we are having our own personal conversations. The book contains 36 interviews — 18 with Nobel laureates — including such notables as Glenn Seaborg, Ronald Gillespie, Gilbert Stork, Alfred Bader, Jean-Marie Lehn, Ad Bax, Ilya Prigogine, Mildred Cohn, and Richard Zare. Each entry includes a biographical sketch followed by a question-and-answer section conducted by Hargittai. Most of the interviews took place in the late 1990s.

The questions probe the early lives of the chemists, the reasons they chose their profession, and how they decided which research problems to investigate. The answers reveal the chemists' aspirations as well as their hardships and triumphs. The book constitutes a valuable historical document that all chemists will find useful, especially for the classroom.

The first volume of the *Candid Science* series, published in 2000, was subtitled *Conversations with Chemists*, whereas the second volume, published in 2002, was *Conversations with Biomedical Scientists*. The fourth volume in the series, subtitled *Conversations with Physicists*, is due out later this year.

7

MAGDOLNA HARGITTAI, ISTVAN HARGITTAI, *CANDID SCIENCE IV: CONVERSATIONS WITH FAMOUS PHYSICISTS*. LONDON: IMPERIAL COLLEGE PRESS, 2004

Foreword by Arno Penzias[1]

In this volume, some three dozen physicists offer the reader uniquely personal glimpses into their chosen profession, the Physics they inherited, worked in, and are leaving for posterity. What was Physics like in the 20th century? Scientific progress during these past hundred years has given us an extraordinary legacy — a wide-ranging and comprehensive foundation of understanding, powerful tools and techniques with which to refine that understanding, and practical applications which touch upon every significant aspect of modern life, as we know it.

In its early decades, 20th-century Physics centered upon the collective efforts of a modest number of European scientists. By mid-century, however, the balance had shifted to a large, and growing, scientific establishment in North America. Enriched by refugees, and

---

[1] Arno Penzias (1933–2024), Nobel laureate in Physics.

enlarged by government support, scientific research in Physics flourished in scope and diversity — providing foundations for the nuclear and information revolutions, and catalyzing transformative advances in Astronomy, Biology, Chemistry, and much else besides.

Indeed, so much has happened that we may sometimes forget that all these advances took place within a time scale not much longer than a single human lifetime. This connection with what may seem a far-distant past is reflected in the conversations captured herein — several of which combine first-hand accounts of key advances made as senior researchers, with personal recollections of people and events encountered as students and fledgling scientists.

Experimental physicists generally gain their clearest insights into Nature's workings by studying phenomena under extreme conditions: energies so high that symmetries break, or so low that quantum condensations occur; the nearly perfect vacuum that Nature was once said to abhor, or the ultra-crowded insides of a neutron star; the mass components of the entire Universe, or the rest mass of a single neutrino. In individual terms, this might mean anything from writing proposals for time on some enormous machine to crafting, assembling, and adapting bits and pieces in a single laboratory.

If anything, the pathways to theoretical insights range even further, probing spaces its practitioners can only explain to interested outsiders by use of analogies, if at all. Small wonder then, that success stories in this arena usually include years of formal study at a famous University and apprenticeships with distinguished theoreticians. And yet, some seeming outsiders can gain entry, and even join the front rank of contributors. Consider, for example, the London-based military attaché who selected a graduate school because of its proximity to his "day job". Without a subject matter expert to guide him, he persisted on his chosen line of research — independently rediscovering a series of mistaken conclusions made by earlier workers, before making the discoveries which earned him a place of honor in this select group.

In the end, therefore, the manifest diversity of life experience might seem to thwart attempts at all but the most general categorizations.

In place of a rigid roadmap, the compilers of this volume have wisely allowed this diversity to manifest itself in an artfully threaded series of anecdotal excursions — much as a reader might get from a personal conversation with the Hargittais themselves. Accordingly, chronology and sub-disciplines play secondary roles herein, as the thread of conversation moves through one topic and then on to the next. Instead, more subtle groupings become evident as one conversation often serves to introduce the next. Note, for example, that Mößbauer follows Bahcall, reflecting their common involvement in cosmic neutrinos. Similarly, we find microwave background radiation grouped with particle physics, reflecting its relation to cosmic nucleosynthesis.

Appropriately, the book begins and ends with recollections. The opening conversation centers upon Eugene Wigner's account of people and events in earlier years — a clear link to the past. While superficially similar, David Schoenberg's recollections of Landau and Kapitza serve to turn our attention to the future. In a fitting coda to what has gone before, this final conversation serves to reintroduce those of us in the West to our colleagues in the East. Split apart for most of the past century, these two spheres of science once again share a common community of knowledge — and common hopes for a fruitful future for science and humanity in this new century.

8

BALAZS HARGITTAI, ISTVAN HARGITTAI, *CANDID SCIENCE V: CONVERSATIONS WITH FAMOUS SCIENTISTS*. LONDON: IMPERIAL COLLEGE PRESS, 2004

Foreword by Arvid Carlsson[1]

I feel greatly honored to be given the task to write this Foreword. It has been a great pleasure to go through the fascinating interviews of yet another nearly two scores of scientists by the Hargittais. What struck me this time as much as before is the enormous individual variation of the characters exposed, making every story unique. Whether such variation is peculiar to scientists, or to human beings in general, or perhaps even to other species, is beyond me. However, one can easily identify an important element that all these individuals have in common: curiosity. Again, one can ask if this is something peculiar to scientists. In this case, I feel inclined to answer yes. Admittedly, the exploratory drive rests on a fundamental instinct of profound survival value that all human beings and a great number of other species have in common. In childhood, the response

---

[1]Arvid Carlsson (1923–2018), Nobel laureate in Physiology or Medicine.

to novelty is a lot more dramatic than later in life. As we grow older our curiosity loses some of its intensity, which is perhaps a sign of maturity. Maybe a common feature of scientists is a slow maturation process, at least in this regard.

An interview, like an autobiography, is of course not an impartial statement. As time goes by we tend to remodel our reminiscences, perhaps to make them more palatable for our self-esteem. People involved in one and the same event will thus often describe it and their role in it differently. For the historian, it must therefore be of utmost value to have access to as many personal accounts as possible of a scientific discovery. In this regard autobiographies and interviews are complementary. An advantage of the interview is that the interviewer can bring aspects into focus that the interviewee might otherwise tend to pass by. In any event, it will remain for the historian to scrutinize all relevant documents in order to come as close to the objective "truth" as possible.

It is remarkable how the Nobel Prize has been able to keep its top position over the years. One may wonder why. At the outset the announcement of the Prize must have been astounding, considering its size and its scope. Subsequently, the Nobel Foundation and the institutions involved in the evaluation process have apparently done a sufficiently good job to keep up the reputation. The existence of such a Superprize is, however, not unproblematic. To some extent, this has to do with Alfred Nobel's Testament. First of all, it brings into focus a distinct, prizeworthy discovery. In fact, the discovery should preferably have been made during the year preceding the award, even though earlier discoveries could be taken into account provided their importance were not immediately obvious. These stipulations should perhaps be viewed against Nobel's own astounding discoveries of dynamite and the like, even though it would be unfair to blame the richly gifted Alfred Nobel for simple-mindedness. Nevertheless, the emphasis of a distinct discovery has probably left out a number of outstanding pioneers who have opened up new important fields without necessarily contributing with any specific discovery. It may have been difficult for the various Nobel Committees to deal with this problem in some cases. In any event, it is obvious that the prizeworthy candidates outnumber the laureates and that the actual outcome will often depend on a number of more or less relevant circumstances. It is regrettable that this fact is not always considered enough and that consequently some prizeworthy candidates feel unnecessarily disappointed.

Once again, the Hargittais are to be congratulated on yet another masterful *Candid Science* volume. It will certainly be enjoyed by a great number of enthusiastic readers.

9

ISTVAN HARGITTAI, MAGDOLNA HARGITTAI, *CANDID SCIENCE VI: MORE CONVERSATIONS WITH FAMOUS SCIENTISTS*. LONDON: IMPERIAL COLLEGE PRESS, 2006

Foreword by Paul Nurse[1]

This is the sixth and final volume of a remarkable series of interviews carried out by Magdolna and István Hargittai with some of the most accomplished scientists of the late twentieth and early twenty-first centuries. These volumes provide an unusual insight not only into some extraordinary science but also into the minds of the scientists who carried out that science. Reading these conversations brings you closer to understanding what makes scientists do science and what they feel while they are doing it. They allow the reader to experience the curiosity which drives scientists to find out how the natural world works or said another way, the passion of just wanting to know, the satisfaction of successfully completing the near impossible experiment, the concerns many scientists feel about the implications of their work for society, and the comradeship and

---

[1] Paul Nurse (b. 1949), Nobel laureate in Physiology or Medicine.

collegiality which marks many scientific endeavors. Gaining access to scientific minds working at this level of achievement is difficult but these interviews often get you close.

Why is it important to understand how science and scientists work? The characteristics of science, respect for observation and experimentation, reliance upon refutable hypotheses, consistency across different fields of inquiry, critical skepticism, all contribute to making science the most reliable approach for gaining knowledge about the natural world. Advances in science have spectacularly changed our understanding of the world and of what it is to be human and have also underpinned many new practices and technologies which have improved the quality of human life. This has happened to such an extent that it is now impossible to imagine a society without the benefits that science has provided. Although science has this universal relevance, scientists are trained specialists and are relatively few in number, and there is always a danger that they can easily become remote from the rest of society. The conversations reproduced in this and the earlier volumes help bridge that gap, and we should all be grateful for the vision and fortitude of István Hargittai in chronicling so many of the stories of science and scientists that mark the present age.

***

A quote from the review by Derry W. Jones, University of Bradford, "Scientists with a human face." *Chemistry World*, July 2007, 65 (one page).

> "Relaxed, doubtless, by the deadpan Hargittai sense of humor, these scientists are remarkably revealing about the freedoms and constraints of later 20th-century science. *Candid Science VI* weighs 1.7 kg but is difficult to put down."

10

ISTVAN HARGITTAI, *THE ROAD TO STOCKHOLM: NOBEL PRIZES, SCIENCE, AND SCIENTISTS*. OXFORD: OXFORD UNIVERSITY PRESS, 2002

Foreword by James D. Watson[1]

The Nobel Prize is revered today almost as much as 50 years ago. Its several Nobel-bearing faculty enhanced greatly the esprit of the University of Chicago when I was one of its undergraduates. More important to me, however, was the 1946 awarding of the Nobel Prize in Physiology or Medicine to the 56-year-old American geneticist, Hermann J. Muller. His presence was the reason why I applied to Indiana University for fall 1947 entry to their graduate school. This Southern Indiana school was still in the intellectual backwater, and I would have thought myself down marketing if Muller had not just moved to Indiana. Though I had been told that Indiana also then possessed several young genetics hotshots, their names themselves would not have led to my going there.

As soon as I arrived in Bloomington, however, I discovered that Salvador Luria and Tracy Sonneborn were much more to my liking. They were much younger and did their research on genetics of

---

[1] James D. Watson (b. 1928), Nobel laureate in Physiology or Medicine.

microorganisms, then genetics at its most exciting. In contrast, Muller's research still focused on *Drosophila,* the organism that he used for his Nobel-Prize-winning 1926 demonstration that X-rays induce mutations. Twenty years later, work on the tiny fruit fly had a tired, almost arcane, feeling. In contrast, Muller's course on Advanced Genetics, which I took my first term at Indiana, was an eye-opener. It made by itself my coming to Indiana worthwhile. No one else had lived through so many great moments of genetics, and I eagerly went to every one of his lectures knowing I would learn more how knowledge about heredity unfolded. In going to Indiana, I wanted to get close to the true essence of the gene. Over the next three years, I greatly benefited from the thoughts of an individual who had come to this objective some 35 years earlier.

Seeing Muller in action was equally important in letting me learn firsthand that Nobel Prize bearers are not superhumans. He was very much the seasoned academic, more at ease with the past than with how to move into the future. But I, having no past, had no choice but to gamble my future on a path that did not yet exist. To make my mark, I had to settle on an objective and stick with it until either I or someone else got to the top. Here my Indiana experience gave me the factual knowledge I needed. I had gone there with no interest in DNA, but by the time I left to seek my future in Europe, I wanted to think about nothing else. Being then obsessive in no way guaranteed my way to the Nobel Pantheon, But it sure helped!!

***

Review by Jeremiah A. Barondess, The New York Academy of Medicine. *Lancet* 2002, 360, 577–578. Reprinted with permission from Elsevier.

The Nobel Prizes are just over 100 years old, and are awarded in physics, chemistry, physiology or medicine, literature, peace, and economics. The science prizes are awarded for outstanding discoveries, rather than as recognition of individuals however distinguished. This volume, written by a physical chemist and informed by his own outstanding career as a scientist as well as some 70 interviews with Nobel Laureates, is engaging, rich with anecdote, and full of detail.

István Hargittai provides an engaging account of the dynamics of nomination, lobbying, nationalistic interests, and the attendant secrecy that surrounds the nomination process. We also learn about the vicissitudes of the prizes at different times — for example, during World War II Hitler interdicted the Prizes for Germans on the basis that one had been awarded to a German pacifist a few years before. The geographical distribution of scientists who have won prizes is perhaps not too surprising: there is a disproportionate number of scientists from the USA, the UK, Germany, and Hungary, with only a few winners from Japan, Spain, or the former USSR. Such variation undoubtedly reflects national investment in science, the primacy of science for young people in these countries, and tradition in research that have translated in part into training grounds for young scientists.

Hargittai alludes what he perceives to be some general traits of science laureates — drive and determination are identified as being crucial. Indeed, according to Peter Medawar, humility is not a useful characteristic at this level of scientific pursuit. Generalisations are less easy to make, however, about the process of scientific discovery itself. Hargittai suggests that the diversity of research approaches precludes any simple characterization. With regard to whether there is life after the prize, once more, unsurprisingly, variation is the answer. Some scientists have continued productive research careers, others have gone off in new directions, and a few have apparently found their competitive edge reduced sufficiently to interfere with further productivity.

Of special interest is a chapter about the scientists who might have, but did not win, the prize. Bruce Merrifield characterized the prizes as a lottery, because of the size and quality of the candidate pool. Some discoveries of unquestioned importance were never honored, either because they were premature and therefore not adequately recognized, or because of the death of the investigator. The latter circumstance precluded award of the prize for physiology or medicine to Oswald Avery, Colin MacLeod, and Maclyn McCarty for the discovery of DNA as the genetic substance. Others have, from time to time, been left out of prizes to which they perhaps had a justifiable claim. Sometimes such omissions resulted from the fact that each prize is limited to three recipients, but in other instances, seem simply to be egregious errors — for example, the award for the discovery of insulin went to Frederick Banting and John Macleod, whereas Charles Best was overlooked.

Although The Road to Stockholm is factual and reliable, it has the flavor of a lengthy conversation with an intelligent and engaging friend. In that sense, it is somewhat idiosyncratic, even though engaging. Hargittai recognizes his idiosyncrasy in his citation of a story about Leo Szilard, who, in putting together some materials on the Manhattan Project, remarked that "he was going to write down the facts. Not for publication, just for the information of God. When his colleague commented that God might know the facts, Szilard replied that this might be so, but not 'this version'". Within these constraints, this volume can be recommended as an interesting visit to the world of Nobles with a companion who, like some of the subjects, may have a clinical variant of what he calls "Nobelomania".

***

Review by Andrew Briggs, Professor of Materials, University of Oxford, "Life after death gives birth to a significant prize." *The Times Higher Education Supplement* (London), January 17, 2003. Reproduced with permission from Chris Havergal, News Editor, *Times Higher Education* and from Andrew Briggs.

In 1867 Alfred Nobel patented dynamite. Within three years he was head of the largest explosives cartel in the world, with patents in every industrialized country. At the height of his success, his brother Ludvig died. A French newspaper confused the names and published an obituary of Alfred. Rob Parsons imagined a scenario such as this: "Alfred... sat down with a cup of coffee, read his own obituary and saw what people had made of his life. But he read phrases like 'merchant of death' and 'his fortune was amassed finding new ways to mutilate and kill'. As Nobel held the newspaper in his hands, he vowed that this was not how he should be remembered, and he decided that... his life would be not just successful but significant."

During his lifetime, Nobel used his wealth to encourage and promote the arts, science and peace, but it is for the prizes endowed in his will that Nobel is remembered. The will states: "The whole of my remaining realisable estate shall be dealt with in the following way: the capital, invested in safe securities by my executors, shall constitute a fund, the interest on which shall be annually distributed in the form of prizes to those who, during the preceding year, shall have conferred the greatest benefit on mankind." The interest was to be divided into five parts, apportioned to physics, chemistry, physiology or medicine, literature and peace.

Istvan Hargittai's book arose out of a lecture given in Cambridge — "How to win a Nobel prize". He says he has talked with 70 Nobel laureates and can share his impression of their careers. He writes: "I am not a historian, neither am I a sociologist; I am a scientist. My book illuminates some decisive moments, both of the scientists' careers en route to the Nobel prize and the science itself that is behind some of the contemporary Nobel prizes. It also covers some discoverers for whom the ultimate recognition from Stockholm never materialised."

The writing style is somewhat wooden, and much of the book reads more as a catalogue of facts than as a story, but those facts are fascinating. Chapters include: "Who wins Nobel prizes?", "What turned you to science?", "Is there life after Nobel prizes?" and "Who did not win?". Some of the answers show a pattern, such as the role of schoolteachers and research mentors, but mainly the chapters are an anthology of nuggets, including splendid insights. Lawrence Bragg inspired his students "to concentrate on problems of central importance, to approach them directly, to waste no time on trivialities".

There is a fascinating discussion of changes in research areas. "We can compare it to a master detective getting a fresh case: he sweeps clean his desk in a matter of hours and switches to his new task.

Future Nobel laureates and other great scientists change the subjects and techniques of their research in a great variety of ways. What is common is the willingness and ability to change."

Hargittai emphasizes the watershed nature of Nobel prizes. Because the prizes are so large, and the number who receive them is so small, the difference in outcome between those who are selected and those who are not is out of proportion to differences in contribution. This is exacerbated by the rule that no more than three people can share each prize. This can lead to an almost arbitrary exclusion of individuals who participated in a discovery (often students, such as those involved in the discoveries of pulsars and of the fullerenes). Being close to a Nobel prize but not being awarded one can be a cause of great misery. Jacques Monod commented: "The prize is very good for science and very bad for the scientists."

A biographer of Isambard Kingdom Brunel observed that he had made very little mention of 19th-century politicians. This was because, he explained, no politicians had nearly as much impact on life as the great engineers. Hargittai similarly bemoans how "British and French children have to learn about the respective kings, queens, and presidents, but much less about the great scientists. Our students, our children, the general public, all of us would benefit from knowing a little more about science and how it comes about."

Growing up in Cambridge, I had two classmates at school who were each the son of a Nobel laureate. I later shared digs with the grandson of a laureate. Perhaps this makes it easier for me to benefit from the inspiration of our great scientists without being overawed by the glittering prizes. I have seen how the best of them benefit not only from Nobel's generosity but also from his example of wanting his life to be significant.

***

Review by Mitch Jacoby, "Lives of Nobel Scientists." *Chemical & Engineering News*, September 16, 2002, 80/37, 40. Reproduced with permission from Mitch Jacoby, Executive Editor of *Chemical & Engineering News*.

Have you ever wondered why some great scientists win Nobel Prizes and others don't? Or whether Nobel Laureates tend to fit some sort of profile based on upbringing, personality, and education? Maybe you've thought about changes in lifestyle that come to prize-winners upon being thrust into the Nobel limelight.

If you've ever spent time — even now — dwelling on these sorts of questions, or if you're just hoping to win your own Nobel science prize someday, then *The Road to Stockholm* is required reading.

Istvan Hargittai doesn't quite set out to answer these questions directly in his new book. Rather, by drawing upon a career filled with interviews with some 70 Nobel Laureates, the Hungarian chemist-turned-writer lets readers form their own answers by sharing with them an impressive collection of anecdotes and interesting tidbits about the lives of prize-winners.

Conducted mainly during the 1990s, Hargittai's interviews reveal a world of details about the background, family life, influence of mentors, and other factors that shaped the lives of scientists who won Nobel Prizes. His examination of documents such as those in the Nobel Archives provides readers with a scorecard full of statistics and historical facts and offers a peek at some of the inner workings of the nominating and awarding processes.

In a chapter addressing adversity, Hargittai points out that some prize-winners grew up in relatively privileged and ideal environments — but many did not. For example, Kenneth G. Wilson, who won the 1982 Nobel Prize in Physics, is the son of Harvard University chemistry professor E. Bright Wilson Jr. And D. Carleton Gajdusek, who was awarded the prize in physiology or medicine in 1976, grew up in an intellectual setting where he became acquainted with great scientists and their work early on.

In contrast, Roald Hoffmann's father was killed in 1943, when Hoffmann was five years old. Hoffmann spent the years of World War II hiding with his mother in Poland to avoid concentration camps, and his later childhood continued to be chaotic. For a short while, he attended a Ukrainian school and then switched to a school in Krakow. Later, he studied in a displaced persons camp in Austria, but by the fifth grade, he had to switch again — this time to Munich.

Each change in location brought a change in the language spoken at school. Yet despite the turmoil and tragedy of his youth, Hoffmann, who did not own a book until he was 16 years old, made great contributions to theoretical chemistry and was awarded the Nobel Prize in Chemistry in 1981.

Similar stories about hardship, poverty, or discrimination in the lives of youngsters — which could have destroyed their educational opportunities — are told by other prize-winners, especially those who lived in Europe during World War II or in the U.S. during the Great Depression. Hargittai writes, "I do not believe that handicap is a prerequisite for the success of Nobel Laureates, rather, they prevailed in spite of these handicaps."

When it comes to personalities, Nobel Prize winners come in all colors of the rainbow. Some famous scientists are known for their assertiveness, self-promotion, and high degree of self-confidence. But while those traits can certainly thrust successful researchers toward center stage, where they're sure to be noticed by their peers — and members of the nominating committees — they are hardly prerequisites for winning a Nobel Prize. In fact, some Nobel Laureates are known for their unobtrusive, soft-spoken nature and for shying away from the spotlight.

Murray Gell-Mann, winner of the 1969 Nobel Prize in Physics, is anything but shy and understated. According to Sheldon L. Glashow, also a Nobel Prize-winning physicist (1979), "Gell-Mann knows almost everything about almost everything, and he is not averse to letting you know that he does, and you don't." Similarly, Gilbert Stork of Columbia University found that his Harvard colleague, Robert Burns Woodward, who won the Nobel Prize in Chemistry in 1965, "had an implicit belief that if he did not produce or suggest something, it had no particular importance."

At the other end of the spectrum are Paul A. M. Dirac and John Bardeen. Dirac, who was famous for his modest, quiet manner and for choosing a career in physics because he doubted whether he had the aptitude for electrical engineering, won the physics prize in 1933. Bardeen won two physics Nobel Prizes: in 1956 and 1972. According to Hargittai, the taciturn Dirac once was asked by physics Nobel

Laureate Richard P. Feynman how he felt upon discovering the Dirac equations. Rather than launching into an account of his contributions to science, Dirac humbly replied, "Good." End of conversation.

The day Bardeen and coworkers at Bell Labs discovered transistors — a finding worthy of considerable excitement — all he managed to share with his wife that evening was, "We discovered something today." In much the same way, Bardeen met a colleague in a hallway after making his second prizewinning discovery. The colleague sensed that Bardeen had something to say, yet it took a while before Bardeen spoke up. Finally, he said, "Well, I think we've explained superconductivity."

Based on extensive interviews with Nobel scientists, Hargittai documents wide-ranging sources of motivation that may have led future prize-winners to pursue careers in science. Chemistry Nobel Laureates William N. Lipscomb (1976), Robert F. Curl, Jr. (1996), and Paul D. Boyer (1997) were turned on to science after having received chemistry sets at around age 10. And Paul de Kruif's 1926 book, *Microbe Hunters*, about uncovering nature's secrets, was cited by several scientists as inspiring them to study science.

Influential teachers certainly played a key role in exciting many young students to concentrate on science. An interesting example is Sophie Wolfe, who was the manager of the science stockroom at Abraham Lincoln High School in Brooklyn, N.Y., in the 1930s. Although Wolfe was not a teacher per se, her after-school science clubs and teaching style influenced a number of future prize-winners, including Arthur Kornberg, who won the 1959 Nobel Prize in Physiology or Medicine, and chemistry Nobel Laureates Paul Berg (1980) and Jerome Karle (1985).

The Road to Stockholm is filled with interesting comparisons between scientists who won Nobel Prizes and those who did not and between scientists whose lives were disrupted by Nobel fame and those who tried to carry on with business as usual after standing in the international spotlight. You won't find a universal recipe for winning science Nobel Prizes in the book, but the variety of ingredients in these success stories makes this a flavorful and interesting read.

11

ISTVAN HARGITTAI, *OUR LIVES: ENCOUNTERS OF A SCIENTIST*. BUDAPEST: AKADÉMIAI KIADÓ, 2004

Foreword by Árpád Göncz[1]

Honored Reader,
The book you hold in your hand requires no preface from me, for the author has written his own apologia. It is compounded of the turbulent time of his youth, and of his life's work and achievement; and through it all speak the voices of his parents, grandparents, and friends.

I wish merely to commend to you his story, woven around his encounters with Nobel laureates and other scientists, great and small. Its message is distilled from his life experiences, their tumultuous historical background, and his dedication to his profession. It reflects throughout the spiritual and intellectual setting, in which Hungarian and Jewish identities are indissolubly linked.

The book is deeply personal — a literary creation, and at the same time an authentic document of science history. It presents an unembellished chronicle of a family that was made to endure the privations and horrors of a haunting era in the history of our country. Whether in its multi-layered totality, the message that emerges is one of sorrow

---

[1] Árpád Göncz (1922–2015) former President of Hungary.

and foreboding, or of optimism and resilience of the human spirit, readers will have to decide for themselves, according to their own values, hopes, and fears.

***

A quote from the review by Henryk Eisenberg, Weizmann Institute of Science, Rehovot, "The view from Budapest." *Nature*, November 11, 2004, 432, 150.

> "When asked whether he believed in extraterrestrial beings, physicist Leo Szilard replied that they were already in our midst: they were called Hungarians, and he was one. The implication was that they had colonized our planet. Readers of István Hargittai's *Our Lives* will certainly be left with the impression that they have, or the scientific world, at least."

***

Review by Werner Schroth, University of Halle-Wittenberg, Germany. *Angewandte Chemie International Edition* 2005, 44, 5169–5170. Reproduced with permission from John Wiley & Sons, Inc.

This new book by István Hargittai, Professor of Chemistry at the Budapest University of Technology and Economics, is partly biographical, partly autobiographical, and partly historical. It is entitled *Our Lives*, not "My Life". The work is centered around the author's unusual experiences and his personal encounters, in the context of our recent history.

Consider what has happened in the last hundred years: promises of salvation, usually accompanied by the misuse of power and by dictatorship, and finally the destruction of human lives on an industrial scale; the word Auschwitz appears 27 times in the book's 9-page index, mainly in connection with the fates of individual Jews, while the names of Adolf Hitler and Joseph Stalin both appear 10 times. Altogether this index contains a comprehensive and almost encyclopedic collection of key words from nearly every aspect of our recent history.

It is right that we should still hear or read the testimonies of reliable witnesses of the time, of people who were affected, and of victims, and that we should continue to be reminded of the crimes that were committed. We should not allow a veil to be drawn over those events through the fading of memories, nor let that history come to be seen as distant and irrelevant. The author of this book is well qualified to write on the subject, as a Hungarian-born scientist of Jewish origin. Generations following ours also deserve to have reliable first-hand answers to their questions.

The past does not stand alone in isolation — it also serves to teach us about the present, and that in turn has lessons for the future, insofar as human reason is still able to achieve something useful. What are the memories that still move this author, from the time in 1944 when three-year-old István and his family were deported to a concentration camp?

The two-page preface explains the book's aims clearly. Firstly, it deals with the dilemma between the Jewish and non-Jewish (in this case Hungarian) identities and is thus concerned with a society that has wide variations in traditions and culture within it; ultimately, it deals with a history of suffering that is without parallel. We learn from the book that the suffering of Hungarian Jews escalated dramatically in the final years of World War II.

Secondly, the author wanted to give a true account of his experiences, to take stock of how his own life has turned out, and to bring together in a single book his knowledge about the fates, careers, and views of many witnesses of the times and friends in the world of science.

Hargittai has talked to a large number of prominent scientists, including many Nobel Prize laureates, and has summarized the conversations in this book. There is a wealth of very informative, and often controversial material. And also much more: the passages dealing with topical or historical themes, often of vital scientific interest, are a rich source of information.

The author's experiences are not described in strict chronological order but are instead presented as a collage of more or less independent fragments, each of which is complete in itself. Hargittai links the different associations and memories together by showing how they are related. For example, his father's early tragic death is linked in different ways to questions about his own life and to the lives of friends and close colleagues. However, this style of "associative writing" makes it difficult to follow the sequence of events and their connections in a strict cause-and-effect sense. Consequently, it would not be very rewarding to read straight through the book at one go — instead, one should enjoy it a little at a time, which is how it was written.

It is unnecessary to describe the contents in detail here. It is divided into 19 main chapters, each of which takes the name of a Nobel laureate: Bergström, Nirenberg, Anderson, Sanger, Hoffmann, Hauptman, Hassel, Semenov, Eigen, Elion, Olah, Lederman, Mössbauer, Kroto, Pauling, Merrifield, Watson, Wigner, Yalow. Other Nobel laureates also have roles in the book, although they are not in the list — they include Butenandt (about whom we learn information given by Benno Müller-Hill) and Heisenberg. Some confrontations are described, but they are reported in an honest way, rather than trying to hide them behind an alleged conflict between individuals.

To summarize: Our Lives is a book that one should certainly read. It describes a varied wealth of life experiences and insights from a European neighbor country, of a kind that we were not aware of before. One cannot fail to notice that, here and there, the book expresses resentful feelings arising from the past, but that is understandable, and should not put one off from reading it. Europe is well worth the effort of overcoming such difficulties.

***

Review by Peter Lax, Courant Institute, New York University. *The Mathematical Intelligencer*, 2008, 30/2, 65–66. Reproduced with permission from Marc Strauss, Publishing Director, Mathematics Journals, Springer Nature.

Istvan Hargittai is a distinguished structural chemist. His book, Our Lives: Encounters of a Scientist, is a curious mixture. Each of the 19 chapters starts with a description of a meeting with a Nobel-Prize-winning chemist, physicist, or biologist, from various nations. Half of them are Americans. These are not interviews but brief histories of the laureates; included is a sketch of their personalities and some of the dramatic events in their lives, such as their struggles to obtain an education for themselves. There are clear, nontechnical descriptions of the work by each of the laureates and implications of their discoveries.

But then the author strays from his subject by free association, to describe the work of other scientists, whose work is related to that of the scientist featured in the chapter, or whose lives have been through similar turns. Almost compulsively, the author returns to tragic events in his own life, the death of his father in 1942 in a forced labor battalion on the Russian front, and the deportation in 1944 of the rest of his family for slave labor in Austria.

Hargittai devotes many pages to chronicling antisemitism: a law passed in Hungary after the First World War restricting the number of Jewish students at universities, the increasingly harsh restriction imposed by laws in the late 1930s, and the bestialities of German, Hungarian, and Ukrainian Nazis during the Holocaust. There was plenty of antisemitism in America, too, in the 1930s. Herbert Hauptmann had trouble getting into graduate school, as did his collaborator Jerome Karle, who was told by the Dean of the graduate school at Harvard: "We have enough Jews in Massachusetts; I am not going to add one from New York."

Hargittai himself suffered from discrimination under the communist regime. Since his father had been a lawyer, a bourgeois profession, he was labeled "class alien," and he was not permitted to continue his high school studies. It was only after the revolution of 1956 that this restriction was lifted. Admission to the university was another hurdle; Hargittai was able to overcome it by sheer persever-ance, but being a class alien, he had to pay tuition; for students from the peasant and working class, attendance was free.

Hargittai received part of his education in Moscow; it had a strong impact on him. In later years he visited Moscow frequently. He writes knowledgeably about the Soviet scientific scene, about Kapitza, Landau, their clashes with the system under Stalin, about antisem-itism in the Soviet Union, and the deadening hand of bureaucracy. Although the possibility of studying with Master of Mathematics and Science in Moscow and Leningrad was one of the rare compensations of living under Soviet occupation, relatively few Hungarians took advantage of it. Hatred of the Soviet Union and the harsh living con-ditions discouraged most; Hargittai was an exception. Another excep-tion was the outstanding mathematician Alfred Renyi who had studied in Leningrad. Renyi described how he beat the bureaucracy; when he was getting nowhere with one official, he demanded to speak to his "nachalnik" (boss). When he couldn't convince the nachalnik, he demanded to speak to his nachalnik, and so on until he got to a sufficiently intelligent person.

Although Hargittai has deep interest in symmetry, a highly math-ematical subject, there is not much about mathematics in this book. There are brief discussions of Vera Sós, Paul Turán, and Paul Erdős; the latter had the foresight to get out of Hungary in 1938.

In fact, three of the scientists discussed in this book have made highly original uses of mathematics. 1. Philip Anderson invented "Anderson localization," a deep property of the spectrum of a class of differential operators.

2. Herbert Hauptman and Jerome Karle solved the problem of determining the structure of crystals by X-ray diffraction. One of their key ideas was using the classical inequalities of Otto Toeplitz on the Fourier transform of positive mass distributions.

3. Eugene Wigner had a lifelong love affair with mathematics. He used random matrices to describe complicated molecules and deter-mined the expected distribution of their spectra. This has developed into a major subject, with unexpected connection with other parts of

mathematics. Wigner also is the author of an intriguing philosophical discussion on "The Unreasonable Effectiveness of Mathematics in the Physical Sciences."

A fourth scientist interviewed by Hargittai, the biochemist Marshall Nierenberg, was enormously proud until the end of his life that, when he was a student at Columbia, he had won the Putnam Competition.

I warmly recommend this book to one and all; it is a lively read.

***

Review by Agnes Heller, philosopher, *Olvasónapló 2017–2018*. Budapest: Múlt és Jövő (publisher), 2018, 179–183. Translated from the Hungarian and reproduced with permission from János Kőbányai, Publisher of Múlt és Jövő.

Our Lives is about our contemporaries, in this case about our contemporaries in science, 20th-century science, physics, chemistry, and mathematics. Hargittai's "brothers" are brothers in science, in enthusiasm for discovery, in joy, as well as in the difficulties of work, in failures.

They are science portraits or, rather, portraits of scientists. Hargittai has interviewed several of them and met others in person. Most of them are Nobel laureates, but not all of them, especially not all the Hungarians. He also mentions Vera T. Sós, Paul Turán, and Paul Erdős, and writes more about Michael Polanyi and László Kiss. His friend George Klein is mentioned several times. We always learn what a scientist discovered and how.

I must admit that although the description of the discoveries is not only professional but also clear, I only understood a few of them, although I knew that if I took a little time, I could understand the rest. But I cannot take the time to do so now. Why were these portraits of scientists so interesting to me? Because they were not only about discoveries but also about people. I got to know them, or I can think that I got to know them. Not all of them were clever (I had already learned that from George Klein), but there were some brilliant minds among them in many ways. It is particularly interesting to see how one "figures something out." I had been interested in this since I was 10 when I read a book about Madame Curie and decided to become a chemist (after three months of being sucked away from science by George Lukacs, I never did.) But I have never forgotten that starting out on an unusual path and being stubborn has something to do with talent. Hargittai's many "sibling stories" have proved this. Among those stories, I was perhaps most taken by Harry Kroto and his C60 molecule. (Perhaps, of course, because the author was particularly interested in it.) But I also liked the "provocative Francis Crick."

The history of 20th-century science is one of the basic histories of the 20th century. It is a great story, a success story. But there is another story of the century: the story of disasters, mass murders, and dictatorships. The two are sometimes intertwined.

Hargittai also speaks of cases where the intertwining was far from innocent. In this way, he introduces us to Benno Müller-Hill's book, *Murderous Science*, and reflects on how the past is coming ever closer to the present.

The history of catastrophe and crime in the 20th century is also the narrator's own life story. Hargittai has succeeded in writing a memoir that is in every way a memoir of the century. He includes his own biography among the biographies of the scientists he features. His biography, like the biography of every man, combines several stories.

Sune K. Bergström, a Nobel Prize winner for medicine is the first scientist to be featured in his book. Hargittai begins Bergström's scientific biography: "Many famous scientists are known to have lost their fathers in childhood. It is a sensitive subject for me because when my father was killed as a slave laborer in 1942, I was only fourteen months old." The memory of the young father, murdered when the author was a baby, remains a scar on the soul of the scholar, now in his seventies.

Hargittai was not born in a shell, but very much outside one. Of course, like everyone else, he had to find himself, but he did not have to turn against a mother or foster father. There were more external obstacles than internal ones. Although his father was an "intellectual," his stepfather was a former shop owner, and that made him a so-called "class alien." Therefore, exactly the opposite happened to him as it did to Gábor Révai, who was essentially of the same age.[2] He was not even admitted to the local high school in Orosháza (where they lived). His mother had to fight like a lioness to get him admitted at all (although not in Orosháza). All this was repeated at the university entrance, where Hargittai was rejected with an excellent exam result, for the same reason. Then he had to fight to be admitted despite his "class alien" origin. Later on, in the easing period of the Kádár regime, several opportunities opened up for him, both in Moscow and in Europe and America. He was already a world-renowned scientist and met Nobel laureates. He was known, even by people who lived in seclusion.

His "problems" reoccurred when his brother "defected" to Israel and as a result, he was denied the opportunity to travel abroad. However, as he writes with some irony, this helped him to write his book on symmetry.

Hargittai is one of those who have never for a moment forgotten their Jewishness. There was a time when his Jewishness played a minor role in his thinking, and then more and more. Together with his mother and brother, he was one of the lucky few who were deported to Austria, where most survived. Many of the Nobel laureates of the 20th century were Jewish, and some of them fled from Poland to America, like others, survivors, who never returned to their "unfaithful" home country from the concentration camps but settled in North America. Some did not want to talk about this, others wanted to tell their story. Hargittai gives an accurate account of them.

As I read his book, Hargittai seems a rare harmonious man. He and his wife, Magdi, share a passion for science. I have already

---

[2]Here the reviewer refers to her preceding review in the same volume of Agnes Heller's *Olvasónapló 2017–2018*, pp 173–179. That book was about Gábor Révai's life. Gábor Révai's foster-father "József Révai was a member of the Rákosi-era dictator-quartet."

mentioned another of their books in my reader's diary.[3] He not only writes about the 20th century, but he is also a man of the 20th century.

What will happen in the 21st?

This book was published in 2003, it is time to republish it. Perhaps the author has something to add from the 21st century?

---

[3]Agnes Heller's review of the Hargittais' New York book reproduced later in this volume (pp 255–258).

12

ISTVAN HARGITTAI, *THE MARTIANS OF SCIENCE: FIVE PHYSICISTS WHO CHANGED THE TWENTIETH CENTURY. NEW YORK: OXFORD UNIVERSITY PRESS, 2006*

A quote from the review by Arthur I. Miller, Emeritus Professor of History and Philosophy of Science, University College London, "The Hungarian Martians: Five of Budapest's finest changed the world in the twentieth century." *Nature*, November 30, 2006, 444, 547–548.

> "*The Martians of Science* tells the gripping story of five brilliant and colorful Hungarian scientists — Theodore von Kármán, John von Neumann, Leó Szilárd, Edward Teller, and Eugene Wigner — who had an extraordinary impact on their profession and on world events in the twentieth century."

***

Review by Barton J. Bernstein, Professor of American History, Stanford University, Stanford, California, "The Hungarian influence in US physics" *Physics Today*, May 2007, 63–64. Reproduced with the permission of the American Institute of Physics. This is a joint review

219

with the book of Kati Marton, *The Great Escape: Nine Jews Who Fled Hitler and Changed the World*. New York: Simon and Schuster, 2006.

Aeronautical engineer and physicist Theodore von Kármán (1881–1963); physicists Leo Szilard (1898–1964), Eugene Wigner (1902–1995), and Edward Teller (1908–2003); and mathematician and physicist John von Neumann (1903–1957) were a group of stunningly intelligent émigré scientists in the US. All were born Jewish, were reared in Budapest, received significant higher education in Germany, and faced substantial anti-Semitism in Europe. Because of their originality, they were sometimes playfully treated as though they were from another planet. By about the late 1940s, the five were respectfully and puckishly called Martians, a term that von Kármán himself may have first applied to them.

Today, not many laypeople know how the five Martians helped to shape science and technology in the modern world: von Kármán and his contributions to jet planes and rocketry; Teller, to thermonuclear weapons; von Neumann, to computers; Szilard, to nuclear chain reactions and fission; and Wigner, to nuclear reactors. In various ways, the Martians maintained an emotional distance difficult to bridge but not hard to discern. A biographer who wants to explore the accomplishments and personalities of each faces substantial interpretive problems. When the men are written about collectively, those problems increase.

Various essays and books have already dealt with each Martian. *The Universal Man: Theodore von Kármán's Life in Aeronautics* (Smithsonian Institution Press, 1992) by Michael Gorn is a brief, uncritical biography. *Genius in the Shadows: A Biography of Leo Szilard, the Man Behind the Bomb* (C. Scribner's Sons, 1992) is a thoughtful, admiring account by William Lanouette and Bela Silard, Leo's brother. Wigner has not been the subject of a book-length biography. His published recollections were largely written by an assistant and apparently finished when Wigner had dementia. Von Neumann evoked an unfriendly critical study in Steve Heims's *John von Neumann and Norbert Wiener: From Mathematics to the Technologies of Life and Death* (MIT Press, 1980), praise in Norman

Macrae's *John von Neumann* (Pantheon, 1992), and a detailed volume in Herman Goldstine's *The Computer: From Pascal to von Neumann* (Princeton U. Press, 1972).

Shortly before his death, Teller produced a self-serving volume with his longtime, admiring aide, Judith Shoolery, *Memoirs: A Twentieth Century Journey in Science and Politics* (Perseus, 2001). He has also been the subject of critical studies, most notably by Gregg Herken in *Brotherhood of the Bomb: The Tangled Lives and Loyalties of Robert Oppenheimer, Ernest Lawrence, and Edward Teller* (Henry Holt, 2002; see the review in *Physics Today*, May 2003, page 59) and by journalists Peter Goodchild in *Edward Teller: The Real Dr. Strangelove* (Harvard U. Press, 2004) and William J. Broad in *Teller's War: The Top-Secret Story Behind the Star Wars Deception* (Simon and Schuster, 1992).

Two new books have contributed to the growing literature on the Martians. Kati Marton's *The Great Escape: Nine Jews Who Fled Hitler and Changed the World* discusses four of the scientists (excluding von Kármán) and five other Hungarian émigrés in an often glib but seldom deep volume. István Hargittai's *The Martians of Science: Five Physicists Who Changed the Twentieth Century* explores more deeply the lives and work of all five Martians.

Marton, a best-selling author and broadcast journalist, boldly decided to write a volume on nine émigré Hungarian Jews from Budapest, her native city. By various routes the members escaped Adolf Hitler's anti-Semitic bloodbath and carved out careers in the West. Marton's chosen group, in addition to the four Martians, includes author Arthur Koestler, film producer Alexander Korda, film director Michael Curtiz, and photographers Robert Capa and André Kertész.

Marton, also Jewish, did not discover her heritage until more than 20 years after she immigrated to the U.S. in the late 1950s at about age 8. Her wide-ranging volume is evocative, poignant, and emotionally sensitive. Unfortunately, the book is slim on research, careless about evidence, and riddled with typographical errors. Nonetheless, she tells good stories and often perceptively recognizes the psychological bleakness beneath the surface in many of the émigrés.

Marton's book has the richness of a deeply moral and personal engagement by someone of a later generation who, by studying her subjects, still seems to be trying to come to grips with being Jewish from Budapest and immigrating to the US. Unfortunately, her work is often naive. She uncritically trusts interviews and most memoirs. Strangely, she also avoids using the key collections of papers by Szilard, Wigner, and Teller. And although she uses some materials from von Neumann's files from the Institute for Advanced Study in Princeton, New Jersey, she does not use his papers, which are at the Library of Congress in Washington, DC. Moreover, she misses some important books and articles.

In contrast, Hargittai's *The Martians of Science* is much fuller and better researched. Hargittai is a Hungarian-born chemist and head of the chemistry and chemical engineering department at Budapest University of Technology and Economics. Although he also eschews using archival files, his book is subtle and thoughtful. He is more critical of the scientists — especially Wigner, von Neumann, and Teller. Hargittai has carefully read the published literature and weaves useful, analytical patterns that indicate knowledge of scientific communities while generally approving the relationship of postwar physics, mathematics, and engineering in the U.S. with U.S. cold war politics.

Hargittai chooses not to get too deeply into the science of his five subjects. He does contend very persuasively that von Neumann, had he lived into the era of Nobel Prizes in economics, would have received that prize for his pioneering work in game theory. Hargittai also reasonably thinks that Szilard's various efforts to "save the world" were at least as meritorious as chemist Linus Pauling's and that Szilard also deserved a Nobel Peace Prize, or at least could have shared one. Hargittai respects Teller's work on the science of molecules, and he thinks that Teller might have done enough important scientific work to be a serious contender for Nobel Prizes in physics and in chemistry. Wigner, of course, did win the Nobel Prize in Physics in 1963.

A rather surprising part of Hargittai's book is his thoughtful yet questionably harsh analysis of Wigner. Others have usually regarded

Wigner as a quiet, courteous, and even self-effacing man. But Hargittai, building on the analytical insights of Abraham Pais and Freeman Dyson, proposes a rather negative interpretation of Wigner, of a man often passive-aggressive who seemed to retreat from conversation and intellectual engagement in person but contrived to minimize and ill serve others with indirect rebukes. Hargittai also notes an unsettling event that Marton, because of her inadequate research, apparently overlooked: When Wigner left Europe in the 1930s, he left behind an out-of-wedlock child. Fortunately, despite her Jewish background, she survived the Holocaust.

Perhaps the treatment of the Martians should end not on troubling stories involving Wigner but, on an illuminating, poignant event of a nature, one unknown to both Marton and Hargittai. Shortly before Teller's death in 2003, when he was the last surviving of the five Martians, I saw him as he stood uncomfortably in his black suit for about 20 minutes one hot summer day on the campus of Stanford University. Teller, who supported the Reagan administration's Strategic Defense Initiative, was not far from the conservative Hoover Institute, where he was a senior research fellow. He looked on irritably, waiting for his driver, as dozens of students busily walked, skateboarded, or biked past him. The old, stooped man was apparently of no interest to them. Not one student seemed to recognize the "father" of the hydrogen bomb.

13

ISTVAN HARGITTAI, *THE DNA DOCTOR: CANDID CONVERSATIONS WITH JAMES D. WATSON. SINGAPORE, ETC., WORLD SCIENTIFIC, 2007*

Comment by Aaron Klug[1] (back of the book)

It is a fascinating book, particularly for those of us who either know him, know large elements of the DNA story, or simply want to hear all the opinions and asides of a great discovery of science. It also gives the flavor of a particular period in the history of science, when biology became even more fascinating than physics.

It is a very good read, and I congratulate Istvan Hargittai on a fine work, certainly also useful and important for future generations of historians (or indeed philosophers) of science and scientific discovery.

***

---

[1] Aaron Klug (1926–2018), Nobel laureate in Chemistry.

Comment by Paul Nurse[2] (back of the book)

This book is interesting because of what it tells us about molecular biology, the most important scientific advance of the second half of the twentieth century, and about Jim Watson, major architect of that scientific revolution and an endlessly intriguing iconoclast and scientific commentator.

The Hungarian scientist Istvan Hargittai reports three intimate conversations with Jim Watson, interleaving these exchanges with his own comments. He expands on points raised, summarizes the historical context, and reports further conversations with other scientists who also participated in the major advances of molecular biology. What emerges is another perspective on this great chapter of science, and on the people who brought it about. But, best of all, we learn something more about Jim himself. We glimpse his constantly probing and worrying mind, the way he picks things over, moving backwards and forwards in a conversation which is more with himself than with his interviewer. We see a mind at work that is restless, wrestling with each new problem and always coming up with something insightful and thought-provoking. It is not that Jim is always right, because he is not. However, even when he is wrong and, very occasionally, a touch hare-brained, it is always in an interesting way. But the one characteristic which comes through repeatedly in these informative conversations is Jim's unbounded intellectual courage. He is completely fearless, saying exactly what he thinks. I doubt it if he has ever made a single public utterance that has been influenced by political correctness or the current mores of the day. Read these conversations to see this bold, courageous, and ever necessary thinker at work.

***

Excerpt from a larger article by Hubert Rehm, Editor of *Lab Times*, "The DNA Doctor."
Lab Times 2008, No 3, 26–27.

---

[2]Paul Nurse (b. 1949), Nobel laureate in Physiology or Medicine.

Recently, I read a short, well-written book by the Hungarian chemist Istvan Hargittai, *The DNA Doctor: candid conversations with James D. Watson*. Driven by the urge to discover the ingredients, which make a good researcher, Hargittai has conducted several interviews with famous scientists. Here Hargittai reports on three conversations he and his wife, Magdi, had with Watson in the years 2000 and 2002.

One of the definite messages in Hargittai's book is clear: Watson is not senile! This is in accord with the impression one gets from the long interview Watson had given to the *Sunday Times*. There, one reads, "His comments, however, although seemingly unguarded, are always calculated". This is corroborated by Hargittai: "[...] however shocking he may appear at times, Watson knows exactly what he can say publicly — like fat women have better sex lives — and what he cannot (about politics and religion, for example)."
... Arrogant? Neither!

Is Watson arrogant as Sydney Brenner wrote? Hargittai asked Watson, "What would be your longest ranging impact?" His answer was: "Probably my books. The discovery of DNA was just waiting to be made; it was not a difficult thing; any good chemist should've arrived at the answer pretty fast. Rosalind Franklin was a physical chemist; she really wasn't a complete chemist. Pauling just goofed beyond any reason; it was crazy. Any good chemist should've found the structure of DNA. But *The Double Helix* was probably unlikely to have been written by anyone beside myself."

Is this the answer of an arrogant person? In fact, all of Watson's answers in Hargittai's book have an air of modesty. And please remember, Watson never co-authored papers with his students if he hadn't made a personal contribution — a deeply modest behavior. I am not sure, whether those snapping at his heels, come anywhere close to this degree of modesty.

It is worth noting that Hargittai's book contains several remarks by Watson about Sydney Brenner. One of them is straightforward, "Sydney is not a nice person".

In *The DNA Doctor* Watson speaks on intelligence and genetics. It is difficult to differentiate between intelligence and motivation, he explains, and it is difficult to define intelligence without proper knowledge of how the brain works. This is in accord with my interpretation of his statement to the *Sunday Times*: If it is difficult to define intelligence — which is true — it is also difficult to make a statement about it. All you can do is bring up trivial assertions like "intelligence traits in genetically different populations are likely to be different since the brain is a delicate organ whose functions are influenced by a lot of genes".

A major trait of Watson's personality seems to be his loathing of hypocrisy, especially when it takes the form of political correctness. "Today it is religiously important to be politically correct", he told Hargittai. Moreover, Watson provokes with a sense of conviction, "If you move toward the center, you are not serving your function as an intellectual".

I believe, Watson made his remarks not out of senility or arrogance but out of contempt for political correctness and the underlying cowardliness. He was the right man to do so. Who else but Watson, a rich man with no further ambitions for personal gain, can be so outspoken against the powerful? Watson seems to have decided that it is more important to fight the "thought police" than to issue careful scientific statements.

A doubtful apology

That's not a problem, however, one aspect I find difficult to comprehend in Watson's attitude is: If you put up a hypothesis, which provokes people, you cannot just say, I have been misinterpreted and withdraw. Even less so if you offer such a doubtful apology as Watson has done, by saying "I did not mean it" in one sentence and "I cannot understand how I could have said what I am quoted as having said" in the next. You have to discuss the issue; you have to show the evidence upon which your hypothesis is based. Instead of saying "I did not mean it" Watson should have said exactly what he meant.

Hargittai, the author of *The DNA Doctor*, agrees, "The attacks against him have been overblown, and really missed the target. Instead of a candid discussion, any meaningful discussion was aborted or abandoned. His having withdrawn indiscriminately what he had said instead of elaborating, added to the damage."

I recommend Hargittai's book, and I recommend taking a closer look at what Watson said.

My immense respect for Watson has certainly not been swayed.

14

ISTVAN HARGITTAI, *JUDGING EDWARD TELLER: A CLOSER LOOK AT ONE OF THE MOST INFLUENTIAL SCIENTISTS OF THE TWENTIETH CENTURY. AMHERST, NEW YORK: PROMETHEUS BOOKS, 2010*

Foreword by Peter Lax[1]

Sixty years ago, Edward Teller was the center of one of the biggest science public policy controversies — the building of the hydrogen bomb. In the next 40 years, he managed to involve himself in more science–public policy controversies. This full-length biography, a scrupulously fair-minded and evenhanded account, is a timely review of the life and influence of this fascinating scientist.

Teller was one of a remarkable group of Hungarian scientists, affectionately referred to as "the Martians," who made outstanding, world-class contributions to their fields: physics, mathematics, computer design and computational methods, chemistry, and aerodynamics. They made equally important contributions to the defense of the United States and of the free world.

---

[1] Peter D. Lax (b. 1926), Abel laureate in Mathematics.

That so many Hungarians, out of a nation of ten million, became leading scientists of the world calls for an explanation. One is the linguistic isolation of Hungarians in a sea of Slavs in Eastern Europe. There was a general feeling that to maintain their identity they had to achieve extraordinary accomplishments. Another factor was that Teller and the other "Martians" were Jewish, and to overcome anti-Semitic prejudices prevalent in Hungary they had to be special, even among Hungarians.

Hargittai describes Teller as the victim of three exiles. In 1926 at age eighteen, he left Hungary because as a Jew he could not be assured of a university career. He went to Germany, studied in Karlsruhe under the great polymer chemist Herman Mark, then under Werner Heisenberg in Leipzig. Next, he held a postdoc position in Göttingen.

This idyllic arrangement came to an abrupt end in 1933 when the Nazi gang came to power; Teller went on his second exile, ending up in the United States at George Washington University.

Hargittai describes in detail Teller's involvement in nuclear energy, which led to his wartime service at Los Alamos, and his growing preoccupation with thermonuclear weapons — and his clash with J. Robert Oppenheimer. Teller correctly perceived that if the Soviet Union developed thermonuclear weapons and the United States failed, the Soviets would dominate the world, so he devoted all his energies to convincing the US government to embark on a program to build thermonuclear weapons. Then he devoted his energies to the design of these weapons.

Calculations indicated that Teller's original design would fizzle. I was present at a lecture on the subject by George Gamow at Los Alamos, in which he illustrated this conclusion by an experiment. He crumpled a piece of paper, put a piece of petrified wood on top of it, lit the paper and exclaimed, "Look, no ignition!"

Subsequently, Teller devised a new design of an H-bomb that was successfully tested. Other people contributed ideas, but it is entirely appropriate to call Teller the father of the hydrogen bomb.

In 1954, in a sinister conspiracy, Oppenheimer was accused of disloyalty and branded a security risk. Teller gave damaging testimony, and Oppenheimer was deprived of his security clearance. Most (not all) of the physics community turned on Teller, refusing even to shake his hand. Hargittai correctly calls this ostracism Teller's third exile.

After his great success with the hydrogen bomb, Teller's judgment deserted him and he was on the wrong side of a series of important issues, some political and some technical. He vigorously opposed the prohibition of nuclear tests in the atmosphere. He minimized the dangers of radioactive fallout, whose hazards were emphasized by proponents of the test ban. Linus Pauling was their leader; one of the dangers he pointed to was genetic damage due to radiation. In one of the lighter moments of the debate, Teller pointed out that wearing pants increases the temperature of the testes, which can also lead to genetic damage. Therefore, to be consistent, proponents of the ban of atmospheric tests should also insist that men wear kilts.

Teller was wrong to insist on unlimited testing; it would have been inconsistent with asking other nations to refrain from developing nuclear weapons.

Teller was equally wrong in opposing the Strategic Arms Limitation Treaties. He could not imagine that after the death of Stalin, the Soviet Union was changing.

During his presidency, Ronald Reagan announced a program, called Strategic Defense Initiative, to "make nuclear weapons obsolete" through countermeasures. Most defense scientists thought this aim could not be achieved, but Teller thought otherwise. One of his proposals was to use X-ray lasers, another was called "Brilliant Pebbles." Neither was realistic. Some claim that the mere idea of the Strategic Defense Initiative was such a challenge to the Soviet defense establishment that it contributed to the downfall of the Soviet Union. Maybe.

Still, Teller was far from being an arch-conservative. In 1946 he joined Leo Szilard and other physicists to help push through legislation that placed nuclear energy in civilian, rather than military, hands. He strenuously opposed a loyalty oath imposed by trustees of the University of California on the faculty. When his friend Stephen Brunauer was falsely accused by Joe McCarthy of being a security risk, Teller came vigorously to his defense. Moreover, he was against secrecy; he liked to point out that in nuclear weaponry, where secrecy is highest, the Soviets are on par with the United States. On the other hand, in the relatively open field of computers, the Soviets are nowhere.

No other scientist engendered so much emotion as Edward Teller. He had hosts of admirers and an equal number who regarded him as evil personified. Hargittai's aim, admirably achieved, is to set the facts of his life straight.

Teller was a very complicated person; Hargittai says that there were at least three Tellers. We shall not see the likes of them again.

***

Afterword by Richard L. Garwin[2]

Over the years from 1947 until his death, my path and that of Edward Teller crossed many times, joining on occasion, intersecting at large angles, or even running antiparallel. I first met Edward (but didn't call him that at the time) in 1947 when I went to the University of Chicago for graduate work in physics, where I soon became an informal assistant to Enrico Fermi in his laboratory and then a formal PhD student. I received that degree from the Department of Physics in December 1949 and joined the physics faculty where I participated fully in all of the activities until December 1952, when I left to join a new IBM Laboratory at Columbia University where I was to work on liquid helium, superconductors, and many other aspects of physics and technology.

My interactions with Edward were already strong in the context of the weekly faculty seminar of the Institute for Nuclear Studies, which was chaired by Willard F. Libby. As Hargittai tells it well, Fermi always had a major role in these seminars, several times coming prepared to discuss a brand-new theory of high-energy particle collisions or of the origin of cosmic rays that within a few days would be a

---

[2]Richard L. Garwin (b. 1928), National Medal of Science, Presidential Medal of Freedom.

fundamentally new paper in the Physical Review. Teller was full of less significant ideas.

During my first of many summers at Los Alamos in 1950, where I went to work on nuclear weapons, I initially read about Teller's activities at Los Alamos during the war in the weekly progress reports from various groups. I was in a good position to understand his current work at the time — which was a longtime fixation of his — with a few associates, to devise a means for configuring the "classical Super" hydrogen bomb. But calculations showed that the reaction would die out rather than propagate well. I helped Edward Teller by responding to his suggestions about diagnostic means for nuclear explosions, some of which were first implemented in the George shot of the Greenhouse series of nuclear tests in the Pacific, May 9, 1951. But Teller's obsession was the full-scale thermonuclear weapon.

As Hargittai tells it, when I returned to Los Alamos in June 1951, I learned from Edward Teller of the idea of "radiation implosion," revealed in a still-secret publication at Los Alamos of March 9, 1951. Edward wanted an experiment that would leave no doubt that this was a feasible path to a thermonuclear weapon, and I responded on July 25, 1951, with a four-page technical memorandum and a large design sketch of what was to be detonated November 1, 1952, as the MIKE shot at eleven megatons of nuclear yield.

I was unaware at this time that in 1949, still with the goal of the classical Super in mind, Teller had identified me as one of twelve people who should be recruited to work on the Super at Los Alamos, two of the others being Enrico Fermi and Hans Bethe.

Judging Edward Teller is an illuminating volume, from which I learned a great deal, not only concerning Teller but also in regard to philosophy. Hargittai discusses at great length the role of Teller and Ulam in the concept of radiation implosion. In 2003 I had access to a letter from Stanislaw Ulam to John von Neumann dated February 23, 1951, in which Ulam describes his idea of having an "auxiliary bomb" that prepares a main charge through compression by means of the shock wave of the "primary" explosive — a term that was used later. Those of us who knew Stan Ulam understand that he would not have done any of the calculations in the Teller–Ulam joint paper.

Moreover, Teller had at times gone out of his way to claim that he had already told people about the concept of radiation implosion as early as December 1950. Some have asked that if Teller had not already worked it out, how could he respond immediately in the initial conversation with his colleague Ulam that essentially "It wouldn't work," and besides he had a better way — the use of the energy radiated from the nuclear explosion and confined by what came to be termed a "radiation case." Knowing both men and being intimately familiar with the technology of the time, I am persuaded that Teller's years-long immersion in the classical Super blinded him to a concept that he and anyone else could have worked out the consequences of in a few minutes — the utility of compression of the thermonuclear fuel. Lest the untutored in weapon technology say, "Of course," it is necessary to state in Teller's own words of 1979 that he had long had a "theorem" that compression would not help. So, according to Teller, if you couldn't burn thermonuclear fuel at normal liquid density, you couldn't burn it at a thousand times liquid density, because the rate of energy loss would be enhanced by just the same factor as the rate of energy gain.

Only when Teller was about to explain to Ulam why his suggestion was wrong, in order "not to waste a lot of time" with Ulam did he consider committing his theorem for the first time to paper. It dawned on him that he had ignored one crucial fact — that with compression the total energy in the particles per unit volume increased linearly with the compression factor, but the maximum energy in the electromagnetic radiation was dependent on temperature and not on compression.

Edward was stuck in a rut and in my opinion would not have thought of radiation implosion for years, although anyone new to the field, in learning the fundamentals of reaction rates, energy loss, inertial expansion, and the like, had a reasonable probability of coming up with the idea. It had not yet happened, and Hans Bethe proclaimed the invention a "miracle," while the head of the Theoretical Division, Carson Mark, judged that anyone trying to work out the Ulam approach of shock compression would have found that the radiation got there first and was a much more suitable approach.

Hargittai asks why my contribution to the thermonuclear program and to the design of the MIKE shot was unknown to the public. It was not even very well known at Los Alamos. The paper of July 25, 1951, had a small distribution list, and when I spoke at Los Alamos in a classified historical discussion in 1993 I recalled the paper very well; yet only the copy that had been sent to Enrico Fermi, also a consultant in the summer of 1951 at Los Alamos, could be located. The detailed proposal was presented probably in early August 1951 to the appropriate committee at Los Alamos chaired by Bethe at the time, and, as Teller states, was thoroughly criticized and then endorsed and built essentially as I had proposed. It was important what was being achieved, and not who had proposed it, and I had by then sensed what has been a guiding principle in my own life, "You can either get something done or get credit for it, but not both."

In his book, Hargittai questions why it was left for a consultant to integrate the ideas floating around at the time. I tried hard to get the very capable Los Alamos low-temperature physics group to do the preliminary cryogenic design of MIKE, but I was told explicitly that the group was "burned out" because of all of the efforts they had put into the Greenhouse George test, to be shot May 1951, and that they needed to return to doing physics. So I did it myself.

I want to add one fact to the Teller life narrative. Hargittai writes, "In the spring of 1963, President Kennedy decided to award Oppenheimer the Fermi Award, which was handed to him by President Johnson on December 2, 1963, the anniversary of the first nuclear reactor, only days after Kennedy's assassination. Teller had been given the Fermi Award the previous year." My role in this has been untold, because it was known only by the eighteen members of the president's Science Advisory Committee and its executive secretary, Dr. David Z. Beckler. Teller's nomination for the Fermi Award had been approved by the General Advisory Committee (GAC) of the AEC and was forwarded to President Kennedy for his approval.

Naturally, it was transmitted through the president's science adviser at that time, Dr. Jerome B. Wiesner, and was presented as an agenda item for discussion by PSAC.

While there was no dissent that Teller's accomplishments warranted the award, the preponderance of feeling was that the award should not be made because of Teller's role in the Oppenheimer trial. In one of the few political acts of my life, I commented that I abhorred what I judged to be Teller's attack on Oppenheimer, but that I thought that Oppenheimer certainly merited, himself, the Fermi Award and that it could only happen if Teller received it first. The argument apparently carried the day, and the awards were made in successive years to Edward Teller and to Robert Oppenheimer.

Hargittai has meticulously and successfully judged Edward Teller. I came away from reading the manuscript feeling enriched not only in knowledge but in wisdom.

15

ISTVAN HARGITTAI, *DRIVE AND CURIOSITY: WHAT FUELS THE PASSION FOR SCIENCE*. AMHERST, NEW YORK: PROMETHEUS BOOKS, 2011

Foreword by Carl Djerassi[1]

Drive and Curiosity is an unusual book by an unusual author. Istvan Hargittai is a working scientist who is also a prolific author of a special genre of books, which — directly or indirectly — describe the nature of *Homo scientificus*. Most of his conclusions are based on direct interviews conducted in a nonintrusive yet searching manner that instills confidence and frequently leads to disclosures on the part of the interviewee that one would not have expected. I speak from personal knowledge, having gone years ago through such a Hargittai interrogation, which prompted me to search in some of his other books for further examples among famous or notorious scientists of confessions, mea culpas, instances of braggadocio, and on occasion even aspects of automythology. The latter is not surprising, since so many autobiographical musings are by definition

---

[1]Carl Djerassi (1923–2015), First Wolf Prize in Chemistry, National Medal of Science, National Medal of Technology.

tainted by automythology as they pass through a person's psychic filter.

In this latest book, Hargittai has collected over a dozen interviews (amplified by literature research) with actual or quasi- (i.e., non-anointed) Nobel laureates, which he has woven into an intriguing account of how great scientific discoveries are made. As the title implies, he assigns drive and curiosity the biggest roles, to which I would have added serendipity — given that in several of his examples sheer good luck was the crucial component. Among scientists, curiosity is a given and serendipity an unsolicited gift. But what does drive mean? It is here that the reader will find intriguing examples, ranging from strict lifelong workaholic discipline to working modes that border on dillydallying. How could drive cover such a broad range? Quite simply because in every case ambition is the key component — the motivation of so many scientists, but at times also their poison.

Hargittai's book has two remarkable features. In fifteen chapters, he manages to cover an extraordinary range of scientific fields — all of them presented in a style that is attractive to readers ranging from sophisticated scientists to laypersons. In the process, he has created a book that may well influence young women and men to select some scientific discipline for their ultimate profession. What makes me say that? In the author's preface, Hargittai states: "From my numerous inquiries, I learned that an inspiring book has turned more children to science than any other single ingredient such as a teacher or a member or friend of the family." I share that opinion and still recall asking that question of the Nobel laureate Joshua Lederberg and hearing him cite Paul de Kruif's *Microbe Hunters* as such a book — a choice that many scientists of my generation (including myself) would have offered, although these days it would seem outmodedly romantic. Apparently, it was also Gertrude Elion's favorite. I would not be surprised if Hargittai's *Drive and Curiosity* will join this list of seminal books.

Hargittai avoids hagiography when describing the work and personalities of the two women and seventeen men that he selected as examples for his generalizations of what made them tick as scientists. With one exception — Kary Mullis — I think that all others illustrate in a positive sense the range of behavioral aspects of *Homo scientificus* from which prospective scientists can learn something constructive. In my opinion, Kary Mullis illustrates none of them except for the operation of serendipity and excessive automythology, coupled with a startling unwillingness to acknowledge the crucial role of colleagues. That the polymerase chain reaction (PCR) is spectacularly important and clearly deserving of a Nobel Prize is unquestionable; I myself nominated him on two occasions for the Nobel Prize, though both times with two collaborators who converted the idea of PCR into practical reality. But much of what is recorded in Chapter 11 about his discovery and his behavior is what I'd consider self-sanitized semi-fiction. Hargittai's description "he appears gentle and diffident, with a sense of self-deprecating humor" may well be true now, but it certainly was not a quarter of a century ago when the discovery was made. The statement "the Nobel Prize paved his way to publish his first and, so far, only book" is also true, but what about that book, titled *Dancing Naked in the Mind Field*? It is essentially a potpourri of autohagiographic ramblings about abductions by outer space aliens, astral travels with a lover, Mullis's belief in the innocence of O. J. Simpson, his conviction of a lack of relation between HIV and AIDS, and his judgment that the impending destruction of the ozone layer is a plot by DuPont rather than the outcome of research by two Nobelists actually featured in chapter 10 of Hargittai's book. I can think of no worse example as a model for a new generation of scientists who ought to be inspired by drive and curiosity.

But the other fourteen chapters more than make up for this controversial pick. Watson, Pauling, Sanger, and probably also Teller are obvious and yet diverse choices that illustrate different aspects of the drive. Neil Bartlett's is a particularly felicitous selection in demonstrating that not every "paradigmatic" discovery ends up with a Nobel Prize. (I liked that Hargittai remembered Primo Levi's understandable error in the first chapter of his *Periodic Table* of assuming that Bartlett had won a Nobel Prize for his discovery that noble gases could also be made reactive.) The inclusion of the two women Nobelists, Gertrude Elion, and Rosalyn Yalow, is especially appropriate since it illustrates the extraordinary hurdles that women faced in science only a few decades ago. Yet instead of glamorizing them solely as heroines, Hargittai shows their strengths, sacrifices, and monumental drive — and in Yalow's case also blemishes.

The inclusion of Rowland and Molina is instructive in many regards, but especially in terms of demonstrating how persistence — another component of drive — can ultimately lead to globally implemented policy changes. Although not explicitly stated, their work on atmospheric ozone depletion is probably one of the best counterarguments to today's chemophobia: by definition, chemicals are the cause of various environmental and toxicological problems, but chemistry is also the answer for combating most of them.

Two Hungarian representatives, Árpád Furka and Leo Szilard, merit special attention. To me, Furka came as a total surprise since I had never heard of him, always having associated his crucial methodological advance first with Bruce Merrifield and subsequently with Mario Geysen. Chapter 6, focusing on Furka, convincingly demonstrates that even scientists suffering under the delusion that they are well-informed on combinatorial chemistry can learn something new and important in this book.

But dedicating an entire chapter to Leo Szilard seemed to me truly inspirational since it also covers an aspect of *Homo scientificus* that is usually ignored, namely, social and even political engagement. Although I have never met Szilard personally, he was always a hero in my mind. Intellectually, he was truly astounding, and Hargittai explains succinctly his enormous scientific contributions in physics and to a lesser extent in biophysics. Yet he won few major scientific awards because he was primarily a planter of intellectual ideas in many fields rather than a consummator in any one of them. But more than most twentieth-century scientists, he understood the global consequences of some of the ideas to which he contributed so deeply — first by pushing for the development of an atomic bomb, but then by dissuading people from using it. He was one of the first participants in the Pugwash movement as well as a founder of the Council for a Livable World, which primarily offers political support to senators who are prepared to concern themselves with arms control (with a focus on small states — the argument being that a senator in a small state, where political campaigns are less costly, has the same vote as one in a large state; after all, only in the Senate are Rhode Island and Texas the same size). I am mentioning these Szilardian activities because Szilard selected a unique method to present and propagate his ideas to a general public, namely in the guise of fiction. In my opinion, his short story *The Voice of the Dolphins* ought to be compulsory reading on literary, humorous, and most importantly political grounds. Briefly, his 1961 short story postulated that during the Cold War, the Americans and Russians decided to collaborate in a neutral site — specifically Vienna — on a project of nonmilitary significance. Founding the Vienna Biological Research Institute with a focus on studying the intelligence of dolphins, they developed AMRUSS — a combined contraceptive and foodstuff, which made the institute financially independent. I shall not monopolize this space so generously offered by Istvan Hargittai to give away the manner in which the institute spent its enormous financial resources in the political arena, other than to mention that it bore an eerie resemblance to actual events of that same period: the founding in 1972 in Laxenburg near Vienna of IIASA (International Institute for Applied Systems Analysis), which used the then-emerging discipline of systems analysis as a medium for peaceful engagement with the Soviet Union. I have often thought that subliminally, Szilard's *The Voice of the Dolphins and Other Stories* eventually influenced me to choose the genre of "science-in-fiction" to propagate some of my own ideas about the behavior and culture of scientists in a belletristic fashion.

The remaining Hargittai choices reach far and wide in terms of subject matter (MRI, X-ray crystallography, conducting polymers, and cosmology), but what I find most impressive is that they — as well as the rest of the book — illustrate the enormous range of work habits of twentieth-century scientists. His summary — the best one can do is doing what one is best at doing — also applies in spades to his contribution as a writer on the nature of *Homo scientificus.*

***

Preface by Harold (Harry) Kroto[2]

Fifteen fascinating scientific discoveries, all but one of which were made in the second half of the twentieth century, are examined in Istvan Hargittai's new book. We learn about the circumstances of the discoveries and also something about the individuals who made them. Although the way the universe works is fundamentally independent of the human spirit, the way that knowledge is uncovered and the way we describe it and use it is not. The mix of knowledge and personal information about individuals and their discoveries that Hargittai presents gives us a deeper understanding of science as a human endeavor and explains why the enthusiasm for uncovering new understanding continues unabated. The book benefits greatly from the fact that not only is the author a practicing scientist with great expertise in a broad portfolio of research areas; he has also been closely associated with all but two of the scientists to whom he introduces us. A further merit is the economical and engaging style of writing, which benefits greatly from his personal knowledge.

The book covers the daring decision of Watson and Crick to embark on a quest, which many judged premature or even impossible, that resulted, eventually, in the elucidation of the human genome. Watson's flamboyant style is contrasted with the natural modesty of Fred Sanger, whose breakthroughs ultimately made the elucidation of the human genome possible. In the story of the development of MRI, we see the struggle for the acquisition of knowledge of the scientists who were responsible for the tremendous benefits MRI now offers as a diagnostic tool. Hargittai introduces us to the additional hurdles that two remarkable women had to overcome on their way to developing new medical approaches. These are but a few of the examples covered in the book. Along the way, he explores the individual variations in the complex mix of curiosity, stubbornness, perseverance, risk-taking, and competitiveness, as well as the other human characteristics that underpin the road to discovery.

---

[2] Harold W. Kroto (1939–2016), Nobel laureate in Chemistry.

Istvan Hargittai has that most valuable quality, in the context of this treatise, a fascination with the personal details involved in important discoveries. His aim is to shed further light on the way scientists work as he uncovers details that are not contained in the original papers and also often not in the personal accounts of the scientists themselves. Because he writes from his own personal perspective and conflates his work with that of the primary authors, we are able to develop a greater overall understanding of the dynamics of the discovery process. This is of value as far as the history of science is concerned, and it is also particularly useful for students who can learn that there are as many ways to do science as there are scientists. In my experience, there is a wide spectrum of general approaches: at one end are the scientists who deliberately set out to solve highly challenging problems, and some are successful — though many, of course, are not; at the other end are scientists who go their own way, exploring problems they find personally curious, oblivious to whether or not the problem seems important to others, or whether there is or is not competition. This latter group, if they do make major breakthroughs, do so unexpectedly and serendipitously. Of course, this cohort includes many who do not make major breakthroughs, but it does tend to include many scientists who are relatively satisfied with their lives and at one with their achievements. The former cohort tends to contain a rather large number of rather disappointed individuals!

Scientists make up less than 0.2 percent of society, even in the developed world, and yet this 0.2 percent is basically responsible for the fundamental discoveries that underpin the technological world we now live in. It is vital that more people understand science not just as a cash cow only to be milked so the industry can make money or even just for the social benefits that might possibly be predicted to accrue. It is vital that the other 99.8 percent of society that benefits from science understands how vital unpredictable and serendipitous discoveries have been and, most importantly, how crucial cultural aspects have been to the science underpinning technological paradigm shifts. The survival of the human race depends on this today more than ever before. Perhaps it is most important that politicians, who daily make decisions on technological issues, understand the most important aspects of the way science works. After all, it should be clear to anyone that though knowledge cannot guarantee good decision making, common sense suggests that wisdom is an unlikely consequence of ignorance.

If this text makes some headway in enhancing the appreciation of science as a cultural activity, it will have done much to improve matters. Hargittai has chosen his examples with care and has astutely covered a broad range of areas from biomedicine to cosmology, and the perpetrators are a richly diverse bunch of characters. In general, the discoveries chosen have profoundly impacted our lives, and the discoverers are always uniquely interesting though perhaps not all are the best of role models. This book is a good read, and scientists and nonscientists will enjoy it and find in it much fascinating, unexpected, and thought-provoking information. Hargittai does indeed achieve his aim, which is to add greatly to our understanding of science as a uniquely human cultural activity.

***

Introduction by Robert F. Curl[3]

Drive and Curiosity shows great science done by individuals of completely different personalities and backgrounds. We learn from it that you can make a huge contribution to the world even if you spent your early career apparently adrift. We learn that you can be filled with passion for your research or apparently be always relaxed and calm and succeed. We learn that great advances in human knowledge often are the result of happy accidents but can also be the result of years of backbreaking toil. We see individuals who come close to the traditional image of the scientist as being shy, wishing to avoid the spotlight belonging to the group. And we find others who are more like rock stars, and still others involved intensely in communication, trying to shape government policy. We follow a scientist trying to make a great discovery where literally the fate of the world hangs upon his success . . . and succeeding.

---

[3]Robert F. Curl (1933–2022), Nobel laureate in Chemistry.

But this book is not just about individuals; it describes how science really works as the vast social enterprise that it is. We see some individuals not getting the credit they deserve until well after their death. We see ethical guidelines crossed by truly great scientists . . . once just to create a joke. We see the dilemma of the young scientist who realizes the advantages of working with a person of great reputation but also realizes that if he happens to be part of a great discovery the community will credit the famous man. We begin to understand how this concern applies with redoubled force if the young scientist happens to be a woman. We become acquainted with women who overcame enormous hurdles to win through to great success and fame.

Drive and Curiosity shows the scientific community punishing individuals because of the notoriety resulting from exposure of private behavior. We see scientists being punished for discovering and reporting unpleasant facts that require a change in the world's behavior. The story of stratospheric ozone depletion is being replayed with a vengeance in the current struggle over global warming. With no easy fix such as that found for ozone depletion, we see scientists being vilified. It has proved much easier to attack the scientist than to attack the science.

Drive and Curiosity depict scientists as real people working in the real world of science, not as unreal people in the fantasy world imagined by the entertainment industry. There inevitably is a lot of science in it. The reader may not be familiar with the science, but it is so clearly explained that the reader is likely to be encouraged to learn more about it. If not, it is a good read anyway. Enjoy!

***

A quote from the review by Derry W. Jones, University of Badford. *Contemporary Physics* 2012, 53, 387–389.

> "*Drive and Curiosity* is recommended as a carefully chosen compelling and enjoyable collection of short biographies and knowledgeable insights into the incentives to discovery of some brilliant minds of the late twentieth century."

16

ISTVAN HARGITTAI, *BURIED GLORY: PORTRAITS OF SOVIET SCIENTISTS.* NEW YORK: OXFORD UNIVERSITY PRESS, 2013

Comments on the back of the book by Richard Garwin[1]

In *Buried Glory*, Istvan Hargittai brings to life 14 outstanding Soviet scientists, and reveals the deadly bureaucracy and terror of the Soviet regime, with imprisonment, murder of family members, and threats being an innate element in their careers. A must-read for anyone with curiosity about our current world, and the one that might have been.

Robert P. Crease[2]

This amazing book is a warm, informed, intimate portrait of what it was like to be a scientist in the Soviet Union, written by an insider who knew many of the subjects. Masterfully written, with unforgettable characters and intricate plot, this book delivers all the pleasures

---

[1] Richard L. Garwin (b. 1928), recipient of the National Medal of Science and the Enrico Fermi Award.

[2] Robert P. Crease, Chairman of the Department of Philosophy, Stony Brook University.

of a Russian novel — except that this tale is true and had a lasting impact on the modern world.

Boris Ya. Zeldovich[3]

An honest, detailed, and breathtaking account of the deeds, ideas, and fates of outstanding scientists of the former Soviet Union. I wholeheartedly recommend it to a broad readership interested in the history of human accomplishment.

Alexey Semenov[4]

This book introduces a unique constellation of brilliant Soviet scientists, and they are described well. The heroes chosen by Istvan Hargittai were exceptional, and their appreciation was high during the cruel and autocratic Soviet period.

---

[3]Boris Ya. Zeldovich (1944–2018) was a member of the USSR (now, Russian) Academy of Sciences, Professor of Optics and Physics, University of Central Florida, son of Yakov B. Zeldovich, one of the main characters of *Buried Glory*.

[4]Alexey Semenov, Professor of Biochemistry and Biophysics, Moscow, grandson of two of the main characters of *Buried Glory*, Nikolai Semenov and July Khariton.

17

BALAZS HARGITTAI, ISTVAN HARGITTAI, *WISDOM OF THE MARTIANS OF SCIENCE: IN THEIR OWN WORDS WITH COMMENTARIES.* SINGAPORE, ETC., WORLD SCIENTIFIC, 2016

Comment by Marina von Neumann Whitman (2016)

No one is better acquainted with the lives of the Martians of Science — five Hungarian geniuses who put their extraordinary talents to work in the cause of democracy and freedom — than the father-and-son team of Istvan and Balazs Hargittai. With widely varying personalities, from the imperious Edward Teller to the courteous, soft-spoken Eugene Wigner, the five had in common, in addition to their incredible brains, a shared history as émigré Hungarian Jews who made new lives in the United States, and whose contributions in the fields of physics, rocketry, and computer development played a critical role in their adopted country's victory in World War II. Much has been written about these men, but only the Hargittais have allowed us a glimpse into their souls through their own words, put in context by the authors. I am the daughter of one of the Martians, John von Neumann, and knew several of the others personally, but this group portrait has given me new insights that I treasure.

***

Review by Mária Vásárhelyi, sociologist, *Magyar Tudomány*, 2016, 177, 1534–1536. Translated from the Hungarian. Reproduced with permission from Mária Vásárhelyi.

This volume represents an unusual and of very high form of science popularization. Balazs and Istvan Hargittai chose to introduce their readers to the most important features of the lives, work, and personalities of the five "Martians" — John von Neumann, Theodore von Kármán, Leo Szilard, Eugene P. Wigner, and Edward Teller. Von Neumann is one of the most famous figures in 20th-century science history, primarily for his pioneering achievements in the field of computer science, but he also played a major role in the development of the hydrogen bomb. Von Kármán played a key role in the development of unified aerodynamics and American aeronautics and is credited with the development of the U.S. Air Force. Szilard made his name in the history of science with his discovery of the concepts of nuclear chain reaction and critical mass and played an important role in the development of the U.S. atomic bomb. The Nobel Prize winner Wigner won the world's most prestigious scientific recognition for his research on the application of symmetry principles to nuclear physics and was also involved in the development of the atomic bomb, as was Teller, who is world famous for his pioneering role in the experimental development of the atomic and hydrogen bomb.

The fates of the five world-renowned scientists of Jewish descent, born and educated in Hungary and later pursuing their scientific careers in the United States, have many similarities. They were all born in Budapest in the fin de siècle, in the last years of the 19th century and the first few years of the 20th century, in upper-middle-class, assimilated, non-observant Jewish families. They were educated in the capital's best high schools under the tutelage of excellent teachers such as László Rátz, the legendary mathematics teacher at the Lutheran High School, whose character we meet in Ferenc Molnár's novel *The Boys of Paul Street*, which has since become a world literary classic. In liberal Budapest, in its heyday, they got to know the world and soaked up the vibrant air of bourgeois life and culture. By the time they had finished their secondary school studies, the happy peacetime had pretty much come to an end. The First World War was followed by the anti-Semitic policies of the Horthy regime and the *numerus clausus* law, the first anti-Jewish legislation in post-World War I Europe. These circumstances forced them all to leave their native country, as it became obvious that they could not pursue an academic career in Hungary because of their Jewish origins. They were correct in their assessment of the opportunities closed to them. During the Horthy era in Hungary not only were the number of Jewish students admitted to universities drastically limited, but university instructors of Jewish origin were not allowed to become professors, and Jews were not elected members of the Hungarian Academy of Sciences.

The five Martians had in common that, after leaving their homeland, they all emigrated to the then-democratic Weimar Germany, where they completed their higher education. Then, when the atmosphere of fascism became unbearable there too, they left for the United States. Each of the five scientists played a decisive role in shaping the destiny of the world, and their scientific achievements radically changed the world we live in.

The five young geniuses who left the European continent in their twenties had further similarities. They knew everything about their own field of science and enriched it with discoveries that would have a world-changing impact. They also remained deeply committed to freedom and democracy and staunch opponents of authoritarian regimes, proving it not only in words and public engagement but also through their scientific work. Except for the engineer von Kármán, all of them were directly or indirectly involved in the development of the atomic and hydrogen bomb. They clearly saw that the democratic world could only protect itself from another devastating world war and the continued expansionist ambitions of dictatorial autocratic regimes by means of such a powerful, beastly weapon.

All five made significant contributions to the universal advancement of science and forever etched their names in the history of science, while at the same time being not only staunch supporters but also staunch defenders of the United States and dedicated their scientific work to the preservation of their new homeland and the free world.

They also shared the common fate of having had to flee their homeland at a young age because of their Jewish origins and the spread of misanthropy, but in their new homeland, the United States, they were given every opportunity to develop their genius and contribute to the progress of science. From the outset, their work was widely recognized socio-professionally and politically and was rewarded with a wide range of professional and political prizes recognition.

Balazs and Istvan Hargittai's aim is that, in addition to the socio-historical and family background and the short biographies of the scientists, the reader should get to know the Martians primarily through their own thoughts and the opinions of their environment. The carefully chosen quotations tell us more about the scientists' thinking, their views of the world and their depth of vision than any careful, accurate interpretation of their ideas might have provided. The fact that we get to know von Neumann, von Kármán, Szilard, Wigner, and Teller primarily through their own words is a major opportunity to get close to their personalities. Through carefully selected quotations, the figures of these scientists are vividly portrayed.

The structure of the chapters on each scientist is similar. In each case, the first step is a brief biography of the 'Martian' to be presented, including the main features of his scientific activity. This is followed by a presentation of the views on the various disciplines and on science in general and the future of scientific activity, history, the world, religion, politics, war and peace, and the use of nuclear energy. A separate chapter is devoted to each scientist to give the reader an idea of the personal qualities and human characteristics of these world-famous Jewish-Hungarian-American scientists.

In addition to the expertly selected quotations, the book's great strength lies in its wealth of illustrations. Photographs and copies of documents on the lives of the scientists bring the reader closer to the characters in the book. At the same time, the photographs in the book, most of which are in black and white, give a good sense of the era and the environment in which the people concerned lived. A particularly interesting feature of the pictures is that many of them show the Martians in the company of prominent figures in American scientific and political life, which is clear evidence that the scientists who fled Hungary were fully accepted participants in the highest circles of political and scientific life in the United States.

It is particularly worth mentioning the very high quality of the workmanship of the volume.

Another important virtue of the book is its clear language, which is easy to understand for the layman. The authors have presumably made a conscious effort to make the book an enjoyable read even for readers who have no understanding of the natural sciences. They carefully avoid scientific jargon and present their characters in an excellent style. At the same time, the book is also an excellent read for those who are at home in the world of physics, mathematics, and science in general, as they can learn about a brilliant representative of their own discipline from a side of the field that may have been unknown to them in their professional work.

18

ISTVAN HARGITTAI, MAGDOLNA HARGITTAI, *NEW YORK SCIENTIFIC: A CULTURE OF INQUIRY, KNOWLEDGE, AND LEARNING.* OXFORD: OXFORD UNIVERSITY PRESS, 2017

Review by Agnes Heller, philosopher, *Olvasónapló 2017–2018.* Budapest: Múlt és Jövő (publisher), 2018, 47–51. Translated from the Hungarian and reproduced with permission from János Kőbányai, Publisher of Múlt és Jövő.

My nostalgic trip started with curiosity, continued with learning, and ended with sadness.

Until now, I thought that after twenty-three years, I somehow knew New York. It immediately turned out that I was wrong. The Hargittai couple introduced me to New York that I did not know.

Let me start with the statues of scientists and scientific symbols. The Hargittais photographed all of them, and I recognized some of them from photographs. Some of the statues I had seen, but never looked at. I didn't know what they depicted. At most, they served as signposts, like in Columbus Square or in the courtyard of Columbia University. I was particularly struck by the friezes on the hotel at 56th Street and Seventh Avenue. This is where my husband Feri and I used to go to the swimming pool every day when we lived next door! And I never noticed these friezes! The only sculptures that caught my eye were those of historical figures, and otherwise, I never gave my interest in sculpture to a level below Michelangelo or, at worst, Rodin.

But I learned much more than that. The Hargittais list New York's most distinguished scientists, including countless Nobel laureates, from 1907 to yesterday. But even from this, I learned that the Nobel laureates of a single New York high school outnumber all the Nobel laureates of Hungary. And we're not just talking about Nobel laureates, we're talking also about inventors that the New World has always boasted, from Benjamin Franklin to Edison to the present day.

The Hargittai couple not only list the natural scientists who came from or lived in New York but also introduce us to their lives, their inventions, their innovations and their discoveries. Here, everyone creates "new" knowledge, something new, something heretofore unknown. The spirit of modern science, a spirit of curiosity and passion, hovers over the city of New York as presented by the Hargittais.

In some places, though very exceptionally, I knew of the discoveries, but not of the discoverers themselves. Sometimes, conversely, I knew the name of the scientist but did not fully understand the essence of what he had discovered. As someone ignorant of the surrounding science and scientific memorabilia, I was given a delicious taster by the Hargittais. There were times when I realized that I had of course "used" one or two of these discoveries, i.e., they had been tried on me, for example, in medicine. It was during the great polio epidemic in Budapest that Sabin drops were tried, and I was given them. Or I never knew (I was not even curious) why the uterine cancer test in New York is called "Pap smear test." From this book, I learned that too.

The Hargittai couple not only showcase the great New York naturalists, they do much more than that. They also get to the why. How was this possible? How did it happen? Where is the secret?

The secret is not to be found in the venerable institutions of the kind found in many big cities. Not in the excellent Natural History Museum or the superb Botanical Gardens. Not even in the New York

State Library, which is truly unique: here, the professor and the homeless man, the university student and the housepainter sit together in the reading room. You can read a Spinoza manuscript and a modern detective novel. But this is not where the secret lies. Where, we can find out from this book.

Above all, in immigration. The proportion of New York Nobel Prize winners in the first and second generations of immigrants is very high. But what made it possible for the sons of immigrant parents who grew up in New York City's poorest neighborhoods and spoke little English to rise to the top from here? Above all, what we call "equal opportunity".

To some extent, "equal opportunity" was not a myth, but a reality. It was deposited in public high schools, where gifted students could get the best education, for free, from the best teachers (Bronx High School of Science, Brooklyn Technical High School, Far Rockaway High School, Townsend Harris High School). The Hargittais not only mention these schools but also tell their stories. Countless Nobel laureates were educated in the New York City Public School System.

These public schools have been linked to several colleges, such as Bronx City College, Brooklyn City College, CUNY (City College of New York), and Hunter College. All offered free tuition even a quarter century ago when I came to New York. Even at Cooper Union, which was a private university, tuition was free.

Of course, it is not enough to provide an excellent public education for the children of poor families. To make good use of this opportunity, you also need the ambition of young people. The ambition of both parents and children. If a person enrolls in a college just to get a diploma, and that's all his parents expect of him, the whole education will be worthless.

Part of this ambition is to ensure that the most talented young people can get into not only good universities but also the best, or rather, the most prestigious, private universities. In New York, in the sciences, this meant Columbia University. Well, getting into Columbia University and rising to the level of researcher or professor was not easy. Especially not for women, Jews, people of color, and immigrants in general. It took decades before all the obstacles were removed.

This is the exact point that saddened me.

This old, substantial, high-quality public (city, state, tuition-free) secondary school and college system is nowhere to be found anymore. Some of them still exist in name, but they are "only" low quality. The teachers are not the primary fault. The students have no ambition other than to get their hands on a piece of paper and get a job. True, there are still two (that, is, very few) high-quality state (public) schools, but these require a strong entrance exam to get in. Young people are sent to private schools by their ambitious parents (if any), who pay a lot for them. As for universities, even in public universities, tuition fees have soared so high that not only the poor but also the middle class cannot afford to send more than one child to school. And the American family is not yet characterized by one child only. Social mobility, the custodian of "equal opportunity," is under threat.

The question is not only where New York's scientists will come from in the future but also where they will go. After all, in today's team-working academic institutions, it is very rare for a single unconventional, pioneering, original talent to find its own way. I am afraid that this book is not only about science but also a history of science. It is as if it were written with an eye to a past that no longer exists.

***

Review by Osmo Pekonen, mathematician, *The Mathematical Intelligencer* 2018, 40/3, 79 (one page). Reproduced with permission from Marc Strauss, Publishing Director, Mathematics Journals, Springer Nature.

István and Magdolna Hargittai, a Hungarian chemist couple, have written, together or separately, an impressive number of books and papers on topics such as symmetry, the interaction of science and art, and biographies of famous scientists. Their latest foray into the cultural history of science is a crunch into the Big Apple: they have produced a richly illustrated guidebook of visible reminders of the presence of science and scientists in New York City. The two authors share a love of the City That Never Sleeps, and they have seemingly roamed every one of its streets — and sometimes climbed into improbable places — to document every statue, plaque, or symbol of famous scientists in the fields of mathematics, natural sciences, medicine, technology, humanities, social sciences, inventions, or explorations. Theirs has been quite an enterprise, to say the least.

As a reminder of the universal character of the metropolis, the Hargittais start their journey from Battery Park, where the monument "The Immigrants" by Luis Sanguino (1983) reminds us of the "huddled masses" that once made the greatness of the city. An Eastern European Jew, a Christian priest, a freed African slave, and an emancipated worker are represented. A Moslem figure could be added at a moment when America, or some part of it, considers the possibility of closing her borders to immigrants from certain countries.

The second illustration is an old family album picture of the Hargittais on top of one of the WTC twin towers that are there no more. The authors point out one very special monument as a symbol of New York's resilience: The disfigured remnants of Fritz Koenig's sculpture "The Sphere" (1971), which formerly graced the plaza at the WTC as a symbol of world peace and harmony, now stands as a tribute to the 9/11 victims in Liberty Park near the site of horror. The monument "now has a different beauty, one I could never imagine," Koenig comments.

The spherical, in the words of Copernicus, is the "form of all forms most perfect, having need of no articulation… the form of the world, the divine body." Our two authors — who as chemists are specialists in symmetry — have identified quite a few spheres, for instance, globes and armillary spheres but also buckyballs, cubes, polyhedra, and other basic geometrical shapes that adorn the buildings, streets, and parks of the great city. This brings us into the realm of mathematics, and I find it wise to limit the rest of the review to that science.

Places of learning with lists of famous alumni are duly presented, with some bias toward Nobel laureates. When browsing the index of names, it seems to me, however, that few mathematicians have got a monument in New York. Josiah Willard Gibbs (1839–1903) and Simon Newcomb (1835–1909) have got their bust in the Hall of Fame for Great Americans, whereas Richard Courant (1888–1972) unsurprisingly has a bust in the institute carrying his name. John Howard Van Amringe (1836–1915), who was one of the founders of the American Mathematical Society, has a conspicuous memorial on the Columbia University campus at Morningside Heights. More subtly, one can find many scientists, among them Archimedes and Pythagoras, depicted in the Gothic ornaments of Riverside Church on the Hudson (Newton, Darwin, and Einstein are there, too). One of the humorous gargoyles at City College of New York seems to be performing a mathematical calculation, so that one may wonder whether some real professor served as a model. Charles Lutwidge Dodgson, alias Lewis Carroll (1832–1898), has been honored in Central Park through a sculptural group by Jose de Creeft (1959) representing Alice's adventures in Wonderland.

As a whole, the New York Community of mathematicians could have done a better job to honor their elders. Surely Martin Gardner (1914–2010) is the kind of figure who should have a monument in the city where he did some of his best work.

The Hargittais don't go as far as inventorying the contents of museums, but many works of art related to mathematics can be found, for instance, in the Metropolitan Museum of Art [1]. They have also written a scientific guide of Budapest, their home city [2].

References

[1] Dauben, J. and Senechal, M. (2015). Math at the Met. *The Mathematical Intelligencer* 37:41–54.
[2] Hargittai, I. and Hargittai, M. (2015). *Budapest Scientific: A Guidebook.* Oxford: Oxford University Press.

19

ISTVAN HARGITTAI, MAGDOLNA HARGITTAI, *MOSCOW SCIENTIFIC: MEMORIALS OF A RESEARCH EMPIRE*. SINGAPORE: WORLD SCIENTIFIC, 2019

Review by György Dalos, historian and author. *Magyar Tudomány*, 2020, 181/1, 132–134. Translated from the Hungarian and reproduced with permission from György Dalos.

When I began my studies at the History Department of Lomonosov University in August 1962, the "physicists–writers" debate was still raging, sparked off by a famous poem of Boris Slutsky.

Physicists and Lyricists[1]

Looks like physicists are in,
looks like lyricists are out.
Nothing Machiavellian, –
nature's law, there is no doubt.

---

[1] Translated from the Russian by Gerald Smith. Source: Boris Slutsky: *Things that Happened*. Moscow: GLAS New Russian Writing, Volume 19, 1998, 251–252. I thank Zsuzsa Hetényi for bringing this translation to my attention.

This the poets have failed to spot,
so-called hypersensitive lot!
Singing scansion and such things
win us only scanty wings,
and the nag that carries us
can't take off like Pegasus.
Looks like physicists are in,
looks like lyricists are out.

This is all self-evident, –
shouldn't cause the least offence.
Arguing's a waste of time.
Best look on with mild attent
at the tidal wave of rhyme
foaming and subsiding, spent,
while the grandeur of its thunder
gravitates
to sign and number.

Slutsky's poem was written in 1958, just as Boris Pasternak's Nobel Prize was causing a worldwide political scandal, to the extent that the poet was forced by the authorities to refuse the Swedish Academy's award. At the traditional banquet in honor of the Nobel laureates at Stockholm City Hall, Pasternak's name was not mentioned, allowing the Soviet ambassador to grace the event without losing face, in order to keep company with three compatriots, Pavel Cherenkov, Igor Tamm, and Ilya Frank, who shared the Nobel Prize in Physics in the same year.

Although my career at that time was directed towards historiography, I saw myself primarily as a poet, which meant that I was on the weaker side of the Slutsky division. Nevertheless, I had a profound devotion to the successes of science and technology, and literature itself led me in that direction, from Dürrenmatt's science fiction physicists, not to mention my communist conviction that in the peaceful competition between the two systems — the slogan Nikita

Khrushchev used to give the Cold War a somewhat sporting character — and in this confrontation, science would play a decisive role. He was excited about who will conquer space first, who will do it better — what a great slogan it was until Chernobyl! — "using nuclear energy for peaceful purposes" and making the endless virgin soil fertile. In my imagination, the nuclear research center in Dubna, the biological institute in Obninsk, but most of all "Akademgorodok" on the outskirts of Novosibirsk, with the "Café Integral" in its middle, were transformed into a veritable socialist polis.

Istvan Hargittai was two years above me at university, and this two-year difference meant that he was an adult compared to me. At that time, apart from my literary glory and the amorous disappointments of my protracted adolescence, I was mostly concerned with the redemption of the world, while in his universe the main focus was on molecular structure research, of which I, of course, had not the faintest idea. Moreover, he read thick textbooks in English, bound together by connoisseurs from the pages of Western publications and distributed as a kind of scientific samizdat. But I considered his greatest achievement to be his correspondence with Eugene P. Wigner, the then Nobel Prize-winning physicist. As a historian, I was also very impressed that he was one of the first to trace the tragic fate of Ervin Bauer, a biophysicist (the brother of the writer Béla Balázs), who disappeared in the Gulag. This endeavor perhaps somewhat anticipated his later work on the history of science.

We had an odd friendship, I knew nothing about his subjects, but he was interested in everything I was interested in at the time. Our passive and active sense of humor brought us closer together, the main subject of which was the discrepancy between Soviet ideology and reality. And although we parted ways for good after graduation, each new casual encounter brought back something of the complicity of student life. On one or two occasions, I had the honor of translating an English or Russian poem for his current work — and I tried to do my best, even though all that remained of my original lyricism was the rhyme and rhythm that Boris Slutsky confronts logarithm half-jokingly, half seriously.

Now, István Hargittai and Magdolna Hargittai's walk through the history of science in Budapest and New York is followed by their third trip, this time to Moscow. For me, this is the most personal for autobiographical reasons. Reading this lavishly researched and illustrated work, I am somewhat retrospectively ashamed of my youthful naivety, which, in the wake of the physicist–writer debate, led me to believe that the real sciences and technology were indeed in a better position than the fine arts. For, on the one hand, it is true that the official cultural policy of the time could more easily find a grip on a modernist poem, an abstract painting, or a dodecaphonic piece of music than on the discovery of a particle, chemical analysis, or a theoretical analysis of numbers, but these concrete sciences, as the countless memories carved in stone show, also suffered the incessant interference of incompetent decision-makers. We are thinking here not only of the tragic fates of certain individuals — Nikolai Timofeyev-Ressovsky, Nikolai Vavilov, Sergei Korolyov — but also of the decades-long taboo on entire scientific disciplines — genetics, cybernetics — and their stigmatization as "bourgeois pseudoscience," which has caused not only moral but also economic damage to the country, amounting to the order of billions. It is understandable that after the "thaw," following Stalin's death, the new generation of Soviet scholars turned with predilection to cultural revival — the "physicists" were eager to read the new output of the "lyricists" — and in Obninsk, Dubna, and the Akademgorods, they welcomed writers and bards excluded from official culture. Nor is it a coincidence that the emblematic figure in the struggle for human rights was a renowned nuclear physicist, Andrei Sakharov.

More important than this is what we learn from the results of the Moscow walks and what we perceive from the images, the indestructible vitality of science, which goes back centuries, which recreates itself under the most extreme conditions, defends its institutions, and fulfills its dual purpose: on the one hand, the direct practice and transmission of its subject and, on the other, the civilizing mission of free thought and innovative action, even in defiance of official authorities.

***

Alexey Semenov,[2] Unpublished statement at the book launch, at the Department of Chemistry, Lomonosov Moscow State University, November 29, 2019. Translated from the Russian and reproduced with permission from Alexey Semenov.

Dear Colleagues!

It is a great pleasure for me to be here today at the launch of Istvan and Magdolna Hargittai's book *Moscow Scientific*. It is eight years since the late Boris Gorobets, the historian of science, introduced us to each other and we have been in constant friendly contact ever since. In 2013, Istvan published his book *Buried Glory*, in which he presented the scientific biographies of 12 outstanding Soviet physicists, chemists, and chemical physicists in a very interesting and objective way.

Now a book on the Moscow monuments of scientists and science has been published. This is of great importance, because hardly any book on the history of science in Russia has been published. Viktor Frenkel, David Holloway, Boris Gorobets, and Gennady Gorelik have written about aspects of Soviet physics and chemical physics and some of their prominent representatives, but no one has written about the monuments of Moscow scientists.

I think that the publication of this book is more than timely and significant at this particular time because we have all witnessed the continuing disregard of scientists in Russia by the authorities in recent years. This includes the decree of Minobrnauki (Ministry of Science and Higher Education of the Russian Federation) regulating the interactions between scientists of Russia and foreign scientists. I am convinced that the narrow-minded policy of the Russian authorities on basic research will lead to a mass exodus of graduates from higher education abroad.

Without basic research, Russia will fall far behind the modern science of developed countries. That is why I consider Istvan and Magdolna's book to be a sincere friendly offering to researchers in Russia, with the aim of drawing attention to the former scientific

---

[2] Alexey Semenov, Professor of Biochemistry and Biophysics, Moscow, grandson of Nikolai Semenov and July Khariton.

greatness of our country. In their book, the authors give a brief but comprehensive and objective characterization of the achievements and creative personalities of modern Moscow science. They have earned our deepest gratitude.

I believe that with this book, Istvan and Magdolna have not exhausted the valuable information in their possession. First of all, I hope that the book will be published in Russian translation.[3] And I also hope that we will be able to welcome other books by the Hargittai couple, including books devoted to science in Russia.

Dear Istvan and Magdi, once again, thank you!

---

[3] The Russian translation of the book from the English was published a year and a half after the English version. И. Харгиттаи, М. Харгиттаи, *Наука в Москве: Мемориалы исследовательской империи* (перевод с английского под редакцией профессора В. М. Тютюнника, Тамбов, Москва, и тд.: МИНЦ "Нобелистика", 2021). The Russian translation faithfully followed the English edition; the only exception was the omission of the critical comments on the government's action of expropriating the network of institutions of the Russian Academy of Sciences. This omission had been agreed upon by correspondence between the authors and the translation editor.

20

ISTVAN HARGITTAI, *MOSAIC OF A SCIENTIFIC LIFE.* SPRINGER NATURE, 2020

Foreword by Agnes Heller[1]

The mosaic of this book is based on meetings. The author, István Hargittai, met with scientists, Nobel laureates, and others, among them friends who were also mostly scientists, during the last decades of the twentieth century and the first of this century. The pattern of the mosaic is alphabetic rather than chronological. The "life" emerging from this mosaic is the author's life. The mosaic tells us who the author is and what he is, and what experiences made him into what he has become; the "who" and the "what" appear in inseparable unison.

Already in his childhood, loyalty, persistence, and curiosity characterized him and these traits have endured for his entire career in science. Thus, for example, first, he had become curious about the occurrences of symmetry, and then he stayed with it and sought connection to those scholars who shared his interest. His persistent dedication then produced books on this topic. His loyalty to his interest did not let him stop at the borders of his research areas; rather, he

---

[1]Agnes Heller (1929–2019), Hungarian philosopher, formerly, Hannah Arendt Professor of Philosophy, the New School in New York.

broadened his inquiries to embrace the great variety of everyday manifestations of symmetry. I was taken by what he had to say about the buckminsterfullerene molecule, and not because I understood exactly what it was, but because I learned that it has been considered to be the most symmetrical and most beautiful among the molecules. What should a philosopher think upon hearing such a designation? That it was almost commonplace for the ancient Greeks that symmetry was the source of beauty, which was first questioned by Plotinus.

Hargittai stresses the same three traits mentioned above, loyalty, persistence, and curiosity in his presentation of scientists — in this book of conversations or a gallery of portraits. For him, it must have been curiosity above all that led him along the path of getting to know so many famous scientists. What else could have lured a renowned scientist, university professor, and recognized author of professional monographs to interview others? What else could have induced him to spend the considerable time necessary for preparing himself, and not only in his own field but in areas that were far from his immediate interest and without expecting them to become his professional interest. This project often involved travels to faraway destinations. Why was he wasting his time? Obviously, he did not consider it a waste. His curiosity — both professional and human — was driving him: to learn about the individual through the accomplishment and about the accomplishment through the individual. This was no journalistic curiosity because the curiosity of the scholar is not directed toward "anything"; rather, it has a specific purpose. This curiosity is aiming at understanding the genesis, the process of becoming — in this case, of becoming a significant scientist.

There are as many roads to becoming a scientist as there are scientists — these paths are impossible to repeat, even for those who already had done it before. The "product," that is, the life of a scientist or a professor, focuses on research or teaching, or both. In addition, the process of the formation of the individual and the individual's role in the world, in history, and the accumulating experience are all also of importance. There must be something common among the individuals who belong to the same world and the same era. The interview, the conversation, is a way to uncover this commonality. The emerging knowledge is not a scientific discovery; rather, it is a discovery in historical anthropology.

Aristotle's Metaphysics begins with the following sentence: "All men by nature desire to know." The question is what and when, and how? How can the desired knowledge be found even if it is never absolute? István Hargittai is seeking the answer to these complex questions in his interviews and other conversations. Max Weber at the start of the twentieth century added that the modern individual should never consider any knowledge absolute, because the next generation will pose new questions from the previous answers. This process never ends, ever. This is what Popper expressed when he said that scientific truth is true only because it is falsifiable.

Despite Weber's skeptical comment, almost all of Hargittai's stories are success stories. These scientists were not looking for "the truth"; they were looking for specific knowledge that was the subject of their research. If they did succeed — and especially if their success was crowned with a Nobel Prize — they could turn elsewhere. Having fulfilled their task, they could change the direction of their inquiry, engage in a hobby, get involved in pedagogy, author books or poetry, do gardening, or embark on world travel. Hargittai values such changes, and he followed such a path as did George Klein, László Bitó, and many others. This already justifies that not all the portraits in the book are scientists. There are others with whom he has been friends, fellow students, or colleagues, or with whom he merely had conversations: from the writer György Dalos to the former president of Hungary, Árpád Göncz. The majority are, however, scientists.

With proper tact, to be sure, but Hargittai considers the traits of the individuals to whom he introduces the reader, and not only their success stories. He has preferences and he does not shy away from letting us know what they are. While recognizing greatness in science, he does not like vanity, jealousy, and careerism, adulation of the powerful, or seeking popularity. He places Crick higher in human values than Watson in the Watson–Crick duo although he equally values

Watson's oeuvre and cannot identify with all of Crick's opinions outside the realm of his science. As for the political profiles of scientists, his stand is unambiguous concerning the services scientists perform for political causes, that is, in connection with the roles they play in defending a cause or display loyalty to one or another.

Hargittai considers his meetings with scientists and their science to be components of the mosaic of his life as they truly are. This is because science has been on center stage all his life. It could not be accidental that his wife, Magdolna, is also a scientist, and they have coauthored several books. It is no mere chance as nothing has been a mere chance in Hargittai's life and neither in the lives of his interlocutors. Chance has played a role in all their lives. A chance did not just stay a chance, because they did something with it. A chance is a good servant if its master knows how to utilize it.

The parallels between Hargittai's life experience and that of his partners play a distinguished role in this book. Hargittai lost his father when he was one year old: He was killed in slave labor. So, when Hargittai meets someone who also lost a father at an early age, he discovers similarities. There are holocaust survivors among his partners, and we learn a great deal about István's and his mother's "lucky" fate as they were deported to a lager (camp) in Austria. When he meets with Jews who had fled Hungary in time, but whose family members were murdered in Auschwitz, we learn about the experience that has impacted the rest of their lives. The activities of émigré scientists who played a determining role in the war efforts of the USA against Nazism, among them the Jewish refugee scientists, are given distinguished attention in the book. They include the five "Martians" (Theodore von Kármán, Leo Szilard, Eugene P. Wigner, John von Neumann, and Edward Teller) about whom Hargittai had written a monograph. From among the émigré scientists, Peter Lax and George A. Olah also figure in separate chapters. Some praise their excellent teachers and Hargittai also remembers his teachers in high school and university, especially those who supported and encouraged him at the critical times when his "class-alien" social status placed him in disadvantaged situations for his studies. These reminiscences remind me the bon mot of my late friend András Pernye. Every significant person had outstanding teachers, he said, and he meant it with irony. Significant people can learn a great deal from almost anybody.

Based on Hargittai's mosaic of his personal and scientific life, I could compose two further patterns. One would be the history of the twentieth century and the other the science history of the same period. The two would not be the same even though the overlaps would be substantial.

Hargittai was a child and youth and became a scholar in Hungary — in a small town and in Budapest. As a persecuted child, he had to wear the yellow star, was condemned to death, and was deported. As a teenager, he was stigmatized with the "class-alien" label and had to struggle for the possibility of studying. This did not crush him; on the contrary, he only became more persistent. Later, still under the Kádár regime, he received a scholarship to continue his university studies at the Lomonosov University in Moscow. Thus, the reader gets some introduction to Moscow and the Soviet Union under Khrushchev and then Brezhnev. He conveys his experience with outstanding Soviet scientists who did everything for protecting their independence and their science. He writes about some with whom he never met, such as the physicist and human rights activist Andrei Sakharov, and others, outside of science, such as the sculptor Ernst Neizvestny whom he met in person only decades later. He remarks about Neizvestny's grave memorial of Khrushchev's tomb that it is an excellent example of antisymmetry. When he mentions Imre Kertész, we learn that *The Union Jack* is his favorite among Kertész's works. I add that even though this is not my favorite Kertész work, *The Union Jack*, is the truest story ever written about the 1956 Revolution.

Then, the reader is introduced to the USA where the author spent so much time in so many places, doing research and teaching, and from whence the largest number of Nobel laureates "came," starting from the 1930s and on. We learn about the school system that, at the time, made it possible even for the poorest, but gifted, child to receive the best education at the highest level.

From these components of the mosaic, the reader can compose the science history of the last half-century. The Nobel Prize, according to Hargittai, does not always reward the most outstanding achievements, and not all outstanding achievements are rewarded with a Nobel Prize — and he refers to the Bible, "many are called but few are chosen." This award plays a different role in the sciences than in literature. One writer is awarded the Nobel Prize in Literature annually and much pain is taken to distribute it properly among continents and languages. In science, there are three Nobel Prizes and up to a total of nine scientists may become recipients. Considerations, such as the mother tongue of the laureate, play no role. In contrast, the recipients of the literature prize are known much beyond a narrow circle of colleagues, whereas this is not the case for most of the science laureates. This was not always so. I have myself been familiar, to the extent a non-specialist might be, with Einstein's or Heisenberg's theories. However, I heard about more than half of the Nobel laureates figuring in this book, for the first time. I have become aware of them only because I read about them in this book, and I learned about their greatness from what they themselves revealed about their dedication, perseverance, loyalty, and insatiable curiosity. I know from one of Hargittai's scientist friends, George Klein, that there are many stupid people even among the Nobel laureates. It seems that Hargittai did not meet them, or he just dreamed his own thoughts into them.

As I said, the reader can compose from these components of Hargittai's mosaic the science history of the last half-century (or even 70 years). This is one of the three patterns (the author's life story, the history, and the science history of Europe and America) emerging here that have overlaps but cannot substitute one for another.

As with all histories, this is also about the past.

It may well be the limitations of my vision, but my impression is that the science history composed from this mosaic is colored by nostalgia. It is as if the author would be saying farewell to the world of science about which he writes. My impression was strengthened by one of his remarks and his reference to one of Arthur Koestler's later books discussing science theory. This is Hargittai's remark: "We live in an era when instant results are expected and there is insufficient tolerance for a universal and inevitably skeptical approach to the big questions of science." This is a clear indication of the diminishing possibility for a paradigm change. In the book referred to by Hargittai, Koestler writes about the role of the unconscious in the great and innovative discoveries and that it often comes from "bisociation," that is, the connection between two independent phenomena linked only by the thoughts of the discoverer. Koestler supports his proposition with numerous examples and testimonials. What is the prerequisite for a great and innovative discovery to happen? For this, individual scientists need to attack the tasks they consider exciting individually: a team of researchers have no collective subconscious. Also needed is time; deadlines cannot produce discoveries. Needed also is a broader approach and an extended thought process to lead to results of greater significance rather than to small additions to the already existing knowledge. If I interpret it correctly, Hargittai's feeling of nostalgia originates from the fast-disappearing conditions for the kind of research that might produce even paradigm changes. I must add that this situation characterizes not only the "natural" sciences, but all bureaucratizing sciences.

I have composed three patterns; some readers might be able to produce a fourth. Try it!

21

ISTVAN HARGITTAI, MAGDOLNA HARGITTAI, *SCIENCE IN LONDON: A GUIDE TO MEMORIALS.* SPRINGER NATURE, 2021

Foreword by Paul Nurse[1]

London is rightly well known as a city of culture, business, politics, and the performing arts, but what is not so well appreciated is that London is also a city of science, in fact, one of the great cities of science in the world. This is a book that celebrates the science and scientists over the ages who have had links to London through 'memorials', a wide range of artifacts and buildings including statues, portraits, plaques, buildings, museums, and scientific research institutions. These reflect the achievements and advances in science not only from those who came from or worked in the British Isles but as befits a great global city, also many from around the rest of the world. The memorials span four centuries ranging from Bacon and Newton in the seventeenth century through to Crick and Dirac in the twentieth.

The breadth of science reflected in these London memorials is quite extraordinary. The compass of science is drawn widely, covering

---

[1] Paul Nurse (b. 1949), Nobel laureate in Physiology or Medicine.

the natural sciences, the applied sciences of medicine and engineering, as well as explorers who expanded humankind's experience of the natural world. The science and scientists covered also have great historical depth. It would not be an exaggeration to say that the birth of modern science took place in London with the establishment of the Royal Society of London, which was based on philosophical foundations of the seventeenth century also closely linked to London. Collections relevant to science have been assembled in the city's great museums of London, such as the British Museum, the Natural History Museum, the British Library, and the Science Museum. These have engaged people in science from all walks of life and of all ages, including myself when as a child I visited these amazing institutions which introduced me to the wonders of the natural world. London not only has museums telling the stories of science but has world-famous research institutions, many of great historical significance: the Royal Institution, Greenwich Observatory, Kew Gardens, and the great universities of University College, King's College, and Imperial College. Science in London is also to be found in the city's Art Galleries, a major example being the National Portrait Gallery.

I was brought up in London and have worked there several different times during my life. Today I still live and work in the city. Over the years I have walked the streets of London always looking at the city through the prism of science. But despite my familiarity with London and the science of London, I have been astonished by what I have learnt from this book, what was revealed to me that I was not aware of previously. There is still much more for me to discover and for you to discover too. It is a book that will guide you to new places, to new memorials, to new people of science and to see all of this through fresh eyes. It is comprehensive, erudite, and knowledgeable, and a pleasure to read and to have as a guide for all those interested both in London and in science.

***

Review by Tracey Clarke, University College London, *Europhysics News* 2022, 53/2, 36.

"If I have seen further, it is by standing on the shoulders of giants." A famous quote from Isaac Newton, echoed by Einstein centuries later when, asked if he stood on the shoulders of Newton, he replied "No, on the shoulders of Maxwell." But Newton's words deeply offended his contemporary Robert Hooke, who was of small stature and took the words as a personal gibe. The book Science in London delves into the stories of science, such as Newton and Hooke's well-recorded shared animosity, and will appeal to anyone interested in the history of science and/or London. Its historical narrative offers us glimpses into the lives and contributions of hundreds of scientists, natural philosophers, explorers, and engineers; their stories told through their memorials in London. Although greats such as Einstein and Newton are well-represented, equal weighting is given to many more men and women whose contributions to the world we live in are just as profound. Who could imagine life without the electricity and telecommunications enabled by the astonishing achievements of Faraday and Maxwell? Or, particularly resonant today, the vaccines pioneered by Jenner? But *Science in London* goes deeper: showcasing other, more obscure, scientists who contributed to momentous discoveries. A great example is Benjamin Jesty, who deliberately — and successfully — inoculated his family against smallpox by exposing them to material from a cowpox-stricken cow … 20 years before Jenner's vaccine. But Jesty neither published nor publicised his discovery, whereas Jenner did (a fantastic lesson for us scientists today!)

Women are given specific mention, particularly for their contributions to medicine, highlighting the colossal barriers and prejudice they faced in both education and practice. Even Queen Victoria is mentioned: by requesting the new innovation of anesthesia during the birth of her eighth child, she enabled the scriptural precedent that suggested women should feel pain during childbirth to be overcome. Her advocacy changed the prevailing attitude of the time and enabled obstetrical anesthesia to be commonplace, something millions of future women would be extremely grateful for.

Science in London uncovers the human side of science. Lord Kelvin may be best known for his substantial contributions to thermodynamics, but here we learn about his enthusiasm for the new long-distance telegraph technology. He proposed to his future wife — and received her answer — via telegraph. *Science in London* describes Lavoisier becoming embroiled in the politics of the French Revolution and being executed by guillotine, only to be exonerated shortly thereafter. We discover the man responsible for the word "banana," the British botanist/spy who smuggled precious tea plants and knowledge out of China to augment the tea industry in India, and who the first person to perform true alchemy was. From a practical perspective, memorial photos are exhibited extensively throughout the book, and postcodes given for those readers seeking the memorials themselves.

Science in London covers the full swathe of human history, from the philosophers and mathematicians of the ancient era, such as Pythagoras and Archimedes, to scientists still alive today. Nor does the book focus solely on British scientists. Instead, it celebrates London's openness and all those contributors to its position as a city of science, both past and future.

22

ISTVAN HARGITTAI, BALAZS HARGITTAI, *BRILLIANCE IN EXILE: THE DIASPORA OF HUNGARIAN SCIENTISTS FROM JOHN VON NEUMANN TO KATALIN KARIKÓ*. BUDAPEST: CENTRAL EUROPEAN UNIVERSITY PRESS, 2023

Foreword by Ivan T. Berend[1]

This book is about 50 scientists who left Hungary during the hundred years between the late 1800s and the late 1900s. The volume at hand is also a major historical document about emigration, one of the central topics of our age. It is well known that Hungary became a country of permanent emigration during the above-mentioned century, and this trend continued even after the turn of the millennium. In the decade of 2010s, another five percent of the country's population chose to work somewhere in the West and

---

[1] Ivan T. Berend (b. 1930), historian, University of California at Los Angeles, former President of the Hungarian Academy of Sciences.

sent two billion euros home every year to their family members. A great many of them will remain permanently abroad.

Emigration from Hungary is, indeed, continuous, and in certain periods especially significant: The turn of the 19th–20th century, the interwar and immediate post-Second World War years, 1956 but also the 2010s, when — quoting the word of the outstanding interwar Hungarian poet, Attila József — millions of Hungarians "staggered out" ("kitántorgott") of the country. Thomas Mann stated in 1941, "what is the meaning nowadays of such terms as homeland and foreign country, when homeland became foreign and foreign became homeland?" Indeed, for between two and two-and-a-half million Hungarian citizens, their homeland became foreign and they had to look for a foreign homeland where they could escape.

Emigration is a rule in economically less developed, poorer countries where landlessness is widespread, where sufficient jobs do not exist, and living standards are relatively low. Hungary was and is an economically moderately developed country. Both in 1900 and 1913, per capita income level — in nominal value — was hardly more than half (56 percent) of the average West European level. This measure of prosperity further declined during the entire 20th and the beginning of the 21st century: in 1950 it was only 48 percent, in 1990 less than 40 percent, and in 2018 only 29 percent of the average West European per capita GDP. The living standard — since the price level was also lower — was somewhat higher than these numbers would suggest: close to half of the West. Lower income levels, lower living standards, less possibilities in life, create — as demographers call it — a "push effect" to emigrate, especially because the West, only a few hundred kilometers away, created a strong "pull effect."

For scientists, however, the main problem was not their living standard, but their research possibilities and/or the political environment. The unwritten law is that poorer countries spend a much lower percentage of their lower income for research than the richer countries. This is clearly visible from the figures of the 2010s. Germany and Austria spent 3.2, the USA and Denmark 3.1, and Israel 4.9 percent of their GDP for research, compared to Hungary, which spent 1.4, Poland 1.3, Turkey 1.1, Russia 1, Slovakia 0.9, Bulgaria 0.7, and Romania 0.5 percent. Since the per capita GDP of the Eastern countries is only between half and one-quarter of the Western level, a much smaller proportion of their GDP for research equals only a tiny fraction of the Western scientific research money. In other words, scholars in poorer countries had only extremely limited research possibilities compared to their Western colleagues. Research costs have become higher and higher every decade, so insufficient research budgets were one of the greatest "push effects" for them. The often intolerant, even hostile attitudes against their new scholarly ideas and work were just additional elements of the problem.

Indeed, the economic factor was only one of the causes of emigration, and for many who were pushed abroad not even the most important one. I quoted above Thomas Mann's statement from 1941 defining the notion of "foreign homeland." He found one in Switzerland, then in the United States, and then again in Switzerland. He did not emigrate from a less developed country, but from 1933 Germany where Hitler rose to power. Authoritarian regimes hate and purge, degrade, and even kill people, suppress those who oppose the regime or those who belong to certain minorities which are blamed for all the country's troubles. Many scientists were slandered as "enemies of their nations" and fled for political reasons.

In modern Hungarian history, this political "push factor" often became equally important and sometimes more decisive than the economic one. Around the turn of the 19–20th centuries, roughly half of the emigrants from Hungary belonged to disparaged ethnic minorities, such as Slovaks, Romanians, Croats, and others who, altogether, represented half of the population. The ruling Hungarian elite wanted to "magyarize" them, and the school system served this goal. Huge groups of these "second-rate" ethnic citizens left Hungary. Another major factor was a vicious anti-Semitism, clearly expressed by the founding of the Anti-Semitic Party in 1883, and one of Europe's last blood libels in Tiszaeszlár in 1882–1883. This atavistic hatred became a dominant political force in interwar Hungary, accelerated by a lost war, the frustration caused by the peace treaty that cut off two-thirds of the prewar territory, then revolutions and counterrevolutions. A democratic revolution in 1918 was followed by a coup and a short-lived communist regime in 1919, soon bloodily defeated by foreign intervention and a nationalist right-wing counter-revolution. Exalted nationalism and right-wing radicalism became dominant in the interwar regime that equated communism with Jews. In 1920, Hungary became the first country in Europe to enact an anti-Jewish legislation, the so-called *Numerus Clausus* law, limiting the number of Jewish students at universities. Discrimination sharply increased during the 1930s, when the country became Hitler's close ally. Several new anti-Jewish laws destroyed the living possibilities of Jews who represented five percent of the country's population, followed by the Holocaust that killed half of them. Half of the remaining half left the country after the war.

The new historical turn, the Sovietization of Hungary, also played an important role in the postwar emigration trends. Freedom was suppressed again, and a new type of persecution hit all those people who served the previous regime, opposed or criticized the new one, together with the relatively well-to-do people, including peasants with estates larger than 15 hectares. Even small-scale workshops and shops — virtually the entire economy — became state-owned, thus the properties of significant layers of society were expropriated. A new type of first harsh, then, after 1956, a milder form of authoritarian system ruled the country for more than forty years. It collapsed in 1989–1990, when democratic transformation began. However, soon after that, a populist-nationalist type of new autocratic regime emerged in 2010 with Viktor Orbán's leadership.

In other words, Hungary, during its entire troubled modern history was dominated by autocratic systems which discriminated and oppressed certain layers of society. Five dramatic regime changes in one single lifetime caused tremendous social trauma, especially because the new regimes often punished and discriminated against those who served the previous one. All these drove mass emigration during most of the last century up to now.

Among the millions of emigrants, there were tens of thousands of highly educated people, intellectuals, scholars, and scientists. Their number, of course, represents a small minority of emigrants. But their importance was much greater than their numbers. In the entire period, a tremendous brain-drain weakened, sometimes undermined the country.

The relative weakness of the country's economy, the existing limitations of financial possibilities for research, together with the intolerant attitudes of oppressive authoritarian systems repeatedly pushed thousands of young independent-minded people, scholars, and future scholars abroad and in doing so decimated the intellectual and scholarly life of Hungary. This book covers the story of leading scientists. Similar works could also present similar stories of social scientists and artists, such as the social philosophers Karl Polányi and Oszkar Jászi; economic advisors of British governments Nicolas Kaldor and Thomas Balogh; Nobel laureate economist John Charles Harsanyi; Béla Bartók, a pioneer of 20th-century music, world-class conductors George Szell and Eugene Ormandy, the virtuoso violinist Joseph Szigeti, and thousands of others.

Reading this book automatically raises the question: what would have happened to those great talents and scientists if they remained in Hungary? Katalin Karikó, one of the inventors and main heroes of the anti-Covid vaccination, who left Hungary in 1985, and one of the personalities whose story is presented in this book, answered this question: "I would have become a mediocre, disgruntled researcher." This highly interesting volume presents personal stories of emigrated scientists; many of whom faced possible disasters and suffering at the beginning of their careers. But exceptional talent and tremendous hard work in friendly environments led to great success in the end. Their stories provide significant historical lessons. The authors, at the end of the book, offer a brief summary of the lessons: "a society is to lose a great deal from being exclusionist and intolerant… a society is to benefit a great deal from being receptive and tolerant, integrating and welcoming."

NAME INDEX